CONCEPTS FOR MOLECULAR MACHINES

CONCEPTS FOR
MOLECULAR
MACHINES

Jubaraj Bikash Baruah

Indian Institute of Technology Guwahati, India

World Scientific

NEW JERSEY · LONDON · SINGAPORE · BEIJING · SHANGHAI · HONG KONG · TAIPEI · CHENNAI · TOKYO

Published by

World Scientific Publishing Co. Pte. Ltd.
5 Toh Tuck Link, Singapore 596224
USA office: 27 Warren Street, Suite 401-402, Hackensack, NJ 07601
UK office: 57 Shelton Street, Covent Garden, London WC2H 9HE

Library of Congress Cataloging-in-Publication Data
Names: Baruah, Jubaraj Bikash, author.
Title: Concepts for molecular machines / by Jubaraj Bikash Baruah,
    Indian Institute of Technology Guwahati, India.
Description: New Jersey : World Scientific, [2017] | Includes bibliographical references and index.
Identifiers: LCCN 2017024589 | ISBN 9789813223707 (hardcover : alk. paper)
Subjects: LCSH: Molecular machinery. | Macromolecules--Industrial applications.
Classification: LCC TP248.25.M645 B37 2017 | DDC 547/.7--dc23
LC record available at https://lccn.loc.gov/2017024589

British Library Cataloguing-in-Publication Data
A catalogue record for this book is available from the British Library.

Typeset by Stallion Press
Email: enquiries@stallionpress.com

Preface

Molecular machines add an independent domain to science and technology to execute properties related to conventional machines at a molecular level in an integrated manner. Miniaturization of a conventional machine and construction of molecular machine have common points in terms of working with small size but each case has an independent approach and methodology. Many molecular machines are inspired by nature so as to replicate biological functions. Biological molecular machines operate with moving parts that are in nano-dimension and are driven by fuel generated through concerted catalytic reactions. Such process may involve various mechanisms and may be made functional by external stimuli or in-built stimuli. For instance, fundamental aspects realised from Adenosine triphosphate (ATP) synthase molecular machine for synthesise of ATP from ADP in living systems suggests that understanding of molecular level would keep open facts related to activities of living beings as a machine does. Fabrication of functional models is expected to enhance performance of new molecular machines that would be evolving with time. Biological molecular machines being too small, their operational aspects are observed under high resolution microscope but action can be visualised by a microscope attaching micron size attachments. Studying various aspects of physical principles associated in construction and operation of such machines helps to look forward feasibility of new design of molecular machines. Coming days are going to see boom on advent of high performance tools to study microscopic world. This would enable one to have handles to deal with molecular machines at ease. An ambient situation for operation and new ideas for new functional properties from molecular machines will emerge. This book is written with an objective to provide such an impetus to deal with fundamental aspects of molecular machines.

Biography of Author

Jubaraj Bikash Baruah obtained his Ph. D. from Indian Institute of Science, Bangalore in 1989. He carried out a post-doctoral study in Tokyo Institute of Technology, Japan. He served Gauhati University (India) as a lecturer during 1989–2015. From there he moved to join as an Assistant Professor and founding member of Department of Chemistry of Indian Institute of Technology Guwahati in 1995. Presently he holds a highest academic grade professor position in Indian Institute of Technology, Guwahati. In between, he served as Head, Department of Chemistry and Dean, Research and Development of Indian Institute of Technology, Guwahati. His major research interest is on novel properties associated with supramolecular systems. He has published over 250 research articles in reputed journals.

Acknowledgments

A book is initiated with some thoughts and understanding the requirement of a topic, hence I firmly acknowledge the scientific communities that contributed to the growth of the topic written in this book. The theme of this book discusses many important aspects drawn from published articles and books, and so I express my acknowledgements to each and every author and publisher whose publications have made the exercise of writing this book possible.

I thank my wife Helen, son Jnanbikash and niece Maina for their enthusiasm and support while writing this book. I am indebted to the publisher World Scientific and also to editor Ms. Sandhya Devi M.G. for inviting me to publish this book.

Contents

1

Introduction to Molecular Machines

1.1 Preliminary Aspects of a Conventional Machine

A machine is a piece of equipment having several moving parts to perform definite work with the aid of power. Machines have contributed immensely to the growth of human civilization. Manufacturing, aesthetic design of products, swift and safe transportations, instant communications are some aspects which have grown tremendously. The concept of machines started from the start of human life before science was developed. This happened due to necessities for survival and the need to improve the quality of food, cloth, shelter and communication. Machines have changed standard of life by improving economy and ease to get things. A macroscopic machine generally operates on many components acting together in a synchronized manner to transfer motion, with force, or energy from one component to another to act as independent or multiple units. Some machines are just for storing information through acquisition and processing of data and others are to reduce the amount of work required to accomplish a given task.

Necessity to perform work, available technology and commercialisation are the most important motivating points to improve a machine and also to abolish a machine. For example let us look at the changes and progress in development in watch. Before watch came into existence, sun was one of the sources used as time keeper. There was also a time when only the hour indicator in a watch was good enough to provide information on time. Mechanical energy was the main guiding factors. Accuracy and precision of time started to grow as requirement different activities. Digital display has made it one of the important time-keeping facilities, which is followed by inclusion of more parameters related to day to day activities. Through the advent of scientific knowledge, in present day a watch has multiple components of several electronic gazette and is used multiple purposes. Even a simple watch has various components function together which are pivots, wheels, shafts, ratchets, screws, springs etc. When all components are assembled, independent function of each component becomes a supporting activity and function of the watch is like an integral machine which is focused to precisely describe time at a particular moment.

One way to classify machines is based on performing a task. A class of machine called sorting machine differentiate properties relating to size, shape, strength, color, smell etc. Thus there are many sub-headings under which they will fall. Another way to name a machine is by narrating the name of the job it does. Washing machine is such an example which washes cloths. Vending machine to get tea or coffee, voice recording machine such as tape or video recording machine to make a film are some other examples. Another way to explain machine is by describing a process that attracts the attention of a common man, such as by narrating life-saving machines used in hospital, a vehicle as locomotive machine. A curious mind questions on the points on purpose, power and fuel of a machine. There are other ways to describe a machine, for example a machine may be named by fuel used; such as coal engine, petrol engine, diesel engine, electrical machine, solar machine. There are also machines based on the route or path they follow; for example aviating machine, navigating machine, locomotives, terrestrial machines etc. On the other hand, classification also goes by efficiency so as to call a machine as super-fast machine or slow machine. When an integrated job is done by several machines it is customary to describe such units as machinery. Thus, machinery is nothing but a set of unitary machines which can be integrated to perform a job. For example, a locomotive machine is constructed with the help of several machines. United those parts of an integral machine falls under category of machineries needed for locomotive factory. A non-performing or an ideal machine may be referred as equipment; which under suitable conditions can be turned on to perform as machine.

From science and technology point of view systematic and formulated knowledge are essential to improve and understand performances. Understanding on these issues in a quantitative and a methodological way provide means to utilize them to generate higher output. Four factors play pivotal roles to scientifically describe a machine, these are (a) Purpose, (b) Design, (c) Performance and (d) Fuel of a machine.

The basic principle of machine is to get an output and the principle of machine lies on the fact that energy is indiscernible and can be converted from one form of energy to another form. For a machine performing mechanical work two prime physical properties namely force and motion becomes very important. Force in terms of push or pull is required to cause a mechanical movement. Forces operating a machine also occur in pair, involving two objects. These situations can be imagined as one object applying force on another object. As a counter effect an equal amount of opposing force is applied by the other object. Such forces can be contact force arising due to bringing two objects in to contact. It can be a force field created around an area in and around vicinity of the one or both objects. As per the types of work done; exerted forces may be frictional force, compressive force or tensile force.

A simple example of a mechanical machine is a pulley. A simple pulley can be constructed by a putting a rope over a rotatable wheel at a fixed position. By pulling the rope, as the wheel rotates the process helps to lift a large mass (Figure 1.1). Such pulleys are conventionally used to lift water from well when sophisticated machines for the same purpose are not available. Pulley transmits a force which enables the rope to easily lift a weight than the force that would have been required by the rope alone. In other word, pulley acts as a machine to modify applied force on the rope. Hence, input force required to do the work gets reduced. Input force is termed as 'effort force' and output force in such a machine is referred to as 'load force'. Functional and design aspect of the machine is essential to understand working principle of a machine. Design aspect of a machine is also important to increase performance of an existing machine. But, identification of the purpose of a machine helps to define physical quantities to be used as descriptor of performance of a particular machine.

A conventional rotodynamic machine functions with a rotor consisting of a number of vanes or blades. While in operation a relative motion

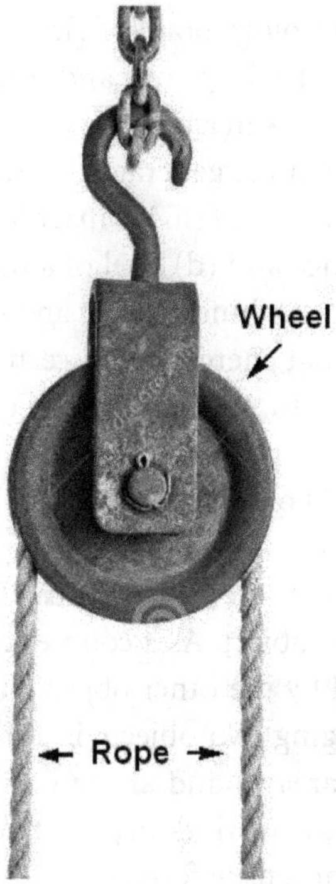

Figure 1.1. An example of a pulley.

between vanes and the fluid medium in gas or liquid occurs. Momentum of the rotor changes when a fluid having definite momentum moves in a direction tangential to rotating blade and passes through the rotor. On the other hand, a still but rotatable blade starts rotating by flow of a fluid. In this process tangential momentum of the fluid is reduced. Similar rationale is applicable in function of pumps and compressors. In these cases the tangential momentum of the fluid increases which helps to extract work done by the fluid from the moving in or out of the pump or compressor.

Let us consider a pump as a machine to illustrate its components and functions. Pump belongs to class of machine used to transfer mechanical energy of a rotor to the energy of fluid. There are several sub-classifications of pumps; for example compressor, fan and blower. A compressor is a pump which compresses gas or fluid. The objective of a compressor is to increase pressure of the gas or fluid under consideration (Figure 1.2). The most

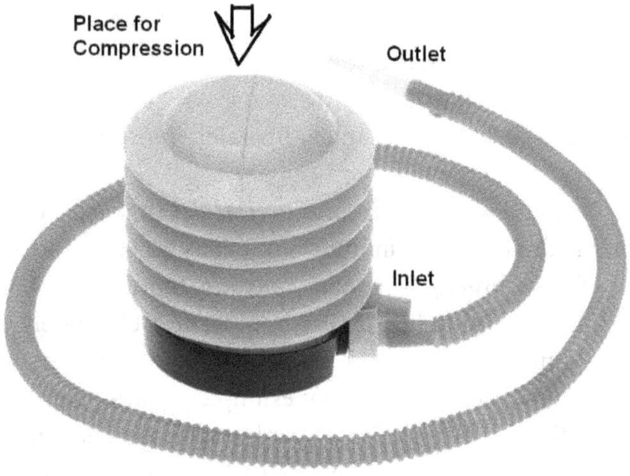

Figure 1.2. A pump to fill balloon with air.

important physical parameter that relates the output is pressure. On the other hand, fan and blower are used to increase or decrease the flow rate of a gas. In these cases mechanical energy of a rotor is used to increase the kinetic energy of the gas/fluid. Thus, main measurable physical quantity as output in this case is kinetic energy. Mathematically it is given by $\frac{1}{2}mv^2$, where m is mass and v is the velocity of the gas/fluid. Thus, in this case mass transfer relates to velocity or acceleration provided to the gas/fluid. Each machine is to give best and suitable output under controlled conditions. For example, a fan has to have control not to cause inconvenience by blowing off the materials in its near vicinity or unless required, pressure exerted by a pump should not lead to an explosion. While utilization of a work it should be possible to regulate the work. Thus, switch and controlling unit becomes essential. To keep a machine in operation external force or stimuli may be required. These forces or stimuli may be acquired from a stored source called as fuel. Alternatively, fuel is required for a machine to operate. Fuel can also be used from a direct source. On the other hand stimuli can be generated from forces such as mechanical, electrical, magnetic, gravitational, light, nuclear forces. Stimuli can also be acquired from another source or diverted form a source which is continuously performing another activity. For example, a pressure can be built by using kinetic energy of flowing water or wind. There are also indirect ways to generate energy and utilize, for example combustion of coal through change of one form of matter to another form. Water transforming to steam is used conventionally in locomotive machines such as in steamers in water transport, balloons in air transport, steam engine of rail etc.

Besides fast and accurate process performed by a machine there are several other issues play roles in improvement as well as utility of an existing version of a machine. For economy purpose the alternative fuel as well as multiple mode of using fuel is very important. Deliberately control on adjustment of input and output of a machine over and above switching on and off, makes a machine versatile. For example, a component called accelerator apparently regulates speed of a car but without a gear it would have very limited performance. Gear has ability to set or modulate ranges of speed that can be altered by accelerator. Thus, gear sets up multiple steps to control and overcome limitation of performance of accelerator. Automation of gear is additional process to sophisticate a locomotive. Embedded computer in such a machine or externally connected computer can be used as the brain of the machine to control all the functions. Further to these, remote control on such locomotive can not only help to keep information on positions and movements in better way but can help in safety measure.

New design, performance, fuel and market economy guides utility of machine utilized in day to day activity. On the other hand machine working with remote controls have great potential and now are routinely used for all kinds of transport systems. But the miniaturisation, integration and automation of machine are universal issues required to make better machine for utilities in a day to day life. Among these requirements miniaturisation has several fundamental basis, as it can be a way to build a machine from molecular understanding to do specific function. Best analogy comes from the function of living beings as each of them perform as integrated machines. In such examples the brain controls output and input. A device or a machine can be very big or very small and their sizes are as per purposes of utilization. Later part of nineteenth century has seen progressive miniaturization of devices and machines has resulted in outstanding technological achievements, especially in the field of information processing.

1.2 Miniaturization of Machines

As mentioned miniaturization of machine is one contention to reduce size and increase efficiency principle.[1-6] At this juncture a question can be framed on the workable or working limit of machinery. The original vision on the miniaturization started with the early work during 1950 by of Richard Feynman. He emphasized the possibility to build machines at nano-meter scale. But the initial studies relating to artificial molecular machines were performed since 1827 by Robert Brown. He was a botanist, who discovered haphazard motions of tiny particles under microscope. Such movement of colloidal particles called Brownian motion was explained with the kinetic theory of matter. Brownian motion of the particles was suggested due to bombardment of the particles by molecules present in a colloidal solution. It was like an imagination at that time due to belief that to build a machine at molecular level a hand to handle of same dimension would be required. With the advent of various optical spectroscopic tools coupled with many other tools relating to diffraction, mass, and techniques suitable for surface analysis the issue of hand being miniaturized has got to a comfortable stage. Feynman had also suggested about bottom up procedure to build machinery. He also initiated theoretical construction, build bottom up concept. In his study he used silicon sprayed layer of atoms by another layer of atoms onto a surface. He studied the effect of partial removal of some such layers. Such removal process created moving parts controlled by electric current, at that time such a construction process was predicted to create optical shutter for tiny camera. With such wonderful predictions, subject has continuously grown and become theme for subject with great future. Synthetic chemists started to deal with such research from about year 1970. However, the subject took acceleration around year 1990 with the development of synthetic processes for interlocked molecular systems which have ability undergo mechanical motion. Slowly control over such movement as well as examples on signal transductions through molecular switches have started to pour in. In recent days, such concepts have developed to prepare improvised molecular machine.

The Nobel Prize in the year 2016 has been shared by three scientists Prof. Ben Feringa, Prof. J. Fraser Stoddart and Prof. Jean-Pierre Sauvage for their contribution toward molecular machines (Figure 1.3).

Sauvage   Stoddart   Feringa

Figure 1.3. Chemistry Noble laureates for molecular machines, awarded in the year 2016.

A machine at macroscopic world functions with Newtonian dynamics and as deterministic models. Dissipation of thermal or any other form of external energy guides the internal motions of molecular machines. The distributions of trajectories are probabilistic in this class of machines.

Energy harvesting from the sun in biology uses molecular devices is also referred to as biological machines. Such processes inspire construction of man-made solar machines. The basic difference is that natural machines are perfect machines evolved over billions of years. Global electronic market now have forced technologists to go for a race towards reducing the size of transistor circuitry with enhanced more powerful microchips. Due to operational advantage at nano-dimension such miniaturization process also has objective to generate new class of nanomotors. Carving graphene into electronic circuits with individual transistors are comparable to the size of conventional molecules. Such transistors are used for printing through a technique called "silver based ink". In this technology, silver nano-particles are used to carry signals and the particles acts as microelectrodes.

Conventional man-made machines in macroscopic world are to be looked at two ways. One way is as assemblies of components fabricated to achieve a specific function. In such machines there are many components performing an individual simple act. While the overall function of the integrated unit is more complex. Each assembled components do characteristic function of the device or machine. In another way, concepts involved in macroscopic machine can be extended to the molecular level. Artificial molecular machines can be benefited by choosing different working principles of natural processes. Nature show functional units causing many work in living beings involves nano-dimensional assemblies, which are yet to be completely understood.

Machine miniaturization has necessitated one to work at molecular level to evolve new ideas for utility. Every machine has its own purpose and utilizes different form of energy. Concept from nanotechnology becomes very important to build minuscule machinery.[7]

One motivating machine that has laid foundation to make molecular machine comes from the electrostatic motor of Benjamin Franklin that was demonstrated in the year 1748. The electrostatic machine designed by Franklin was comprised of small insulated charge containers which were arranged as a rotating conveyors as illustrated in Figure 1.4. This part has a freely rotatable vertical shaft made of several glass bars extended like spokes of a cycle wheel. The ends of these bars projecting outwards were thimble with brass. Two oppositely charged electrodes in the form of Lyden Jars were placed across the wheel. The charge containers could be

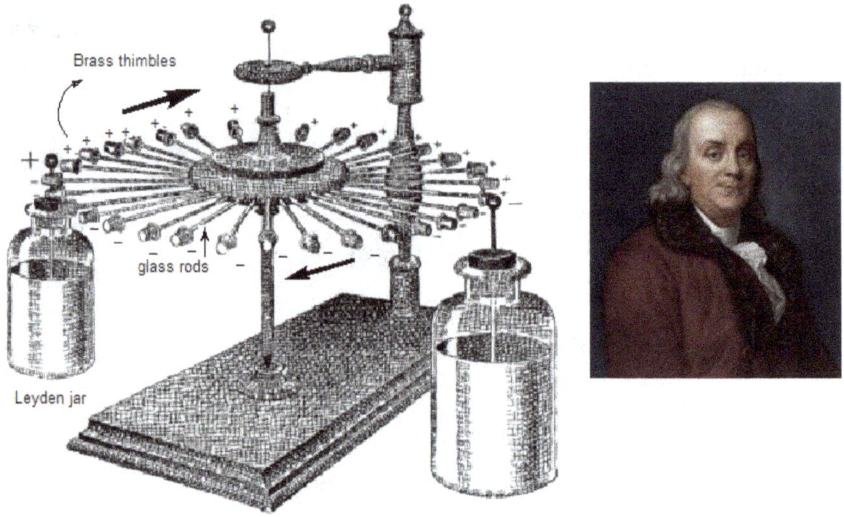

Figure 1.4. Electrostatic motor constructed by Benjamin Franklin (right side is photograph of Benjamin Franklin). (http://ethw.org/Benjamin_Franklin's_Electric_Motor).

approached by two oppositely charged electrodes directly placed in opposite side of the wheel. When charge containers were approached by the
electrodes they got charged. Since electrode of similar charges repel each
other and a rotational motion took place. Reversal of charging or neu-
tralization of the charge on the sticks took place when the charge
containers moved and reached the other electrode, where charging took
place once again. Once charged, repulsive forces made wheel to rotate,
repetition of such process generates circular motion.

This model of machine got tremendously evolved as the magnetic
induction phenomenon was discovered. Present days macroscopic scale
motors uses primarily magnetic effect rather than electrostatic effects.
In miniaturized machines magnetic effects becomes less prominent to
act as source of energy than the electrostatic effects. Due to such a reason it is more convenient to scale up the Franklin's original concept to
microscopic level.

Control provided to a chemical process and structural modifications
by using computer have made it possible to model folding-unfolding of
molecules. Thus the concept of reversible and irreversible knots may be
dynamically modeled to utilize as machine. Many natural folding processes to provide structures such as folded, globular, hair-pin structure of
protein can be modulated to generate signals. This is possible due to
understanding on energetic of transformations between such states.
Artificial molecular machines of biomaterials are designed by utilizing
such changes and by identifying a work to be done by considering different
equilibriums. In present days such information and designs have spread
over a wide spectrum of examples to utilize them for drug delivery, molecular imprinting, electronics and military devices.

1.3 Molecular Machine

The term molecular machine is often used to explain interchangeable
molecular systems. A molecular machine is designed to carry out a function at molecular level. On the other hand a molecular motor is a
component of a molecular machine or independent unit used to convert
energy into mechanical work. Electronic and/or nuclear rearrangements
within a molecule or in an assembly of molecules or Brownian motion

caused by stimuli are used for operation of molecular machines.[8] Similar to conventional machines, molecular machines can be differentiated based on type of (i) function it performs; (ii) energy used to perform work; (ii) motions utilised or generated to do a work, such as translation, oscillation and rotation performed in a singular or combined way; (iv) Based on the means by which the work and performance of the machine is monitored; (v) path or trajectory of the performance such as repeatability of path in cyclic process; and (vi) time at which the machine performs, which may be at nano- or pico-second time-scale.

Food is the fuel to run various activities of a human body. Alternately, a human body uses chemical energy derived from food through many slightly exergonic reactions. After intake of food, digestion is an action to perform work and excretion is a process to maintain waste. Similarly in a molecular machine working on chemical energy sources will have to constantly use new reactant/s. Thus, waste product formation is also a principal factor deciding utilisation aspects of a molecular machine run through or depending on chemical transformations. Similar to macroscopic machines, molecular machines will also require subsidiary part to dispose accumulated waste products. A molecular machine running with chemical reactions has this aspect as disadvantage. Molecular machine utilising photochemical and electrochemical energy as inputs yield endergonic and reversible reactions. This aspect requires understanding of supramolecular photochemistry and electrochemistry. Isomerization and redox processes are two most prominent reactions that is utilised for operation of molecular machines. Use of lasers has provided avenues to work in small spaces with very short life-time to execute a particular operation. Natural motors work without interruption from a routine performance and are autonomous. However, a molecular machine requires additional dependent process to revert back and to perform an operation in the same path; hence they do not tend to reset themselves. Readable changes of chemical or physical property by electronic and/or nuclear rearrangements of a molecular component of a molecular machine is needed to control and monitor their activities. If we consider moving mechanisms in biological systems at molecular level, in vitro techniques that combine optical and mechanical methods to observe the behaviour of a biomolecule becomes a practical part of molecular machines

Molecular machines may be classified by the type of work or by types of stimuli used in such a machine. From chemist point of view, a better way would be to classify them in terms of class of molecules from which such machines are devised. Similar is also the case with bio-related molecular machines, which can be classified based on class of bio-molecules such as protein or DNA-based molecular machines.

1.4 Terminology in Molecular Machines

Machines have wide applications and may also have several components. Hence, there is a need to use consistent terms to provide an effective scientific meaning to describe them by ways in which they perform. Scientific definitions of certain common terms that used in the field of molecular machines are available.[9] Analogous to macroscopic machine definition, a mechanical movement that accomplishes a useful task at molecular level is called a molecular machine. In a molecule or assembly of molecules where stimulus guides motion or movement along a direction one or more molecular component/s relative to another and results in performing a net amount of work are broadly referred to as molecular machines. Such machines are further categorized into two broad classes one is motor and other is switch. For example, a switch may be designed on the basis of the change in position of a ring like structure on the thread of dumb-bell shaped structure and in such cases the operation of movement is just a function of state. Upon retrace of the path to original place nullifies the work done through the initial motion. Hence switches are not suitable for repetitive as well as progressive work. Functional aspect of a motor is in terms of trajectory. Due to this trajectory of a molecular event such that the motor regain original state; a work done during such a path does not get nullified. For example a complete rotation of a rotor along a circle by effecting 360° rotation, work done cannot be nullified unless motor subsequently rotates to make a reverse rotation. Accordingly, a molecular machines falling under category of switch cannot perform progressive work like biological motors does; but they can be a part of such machines to regulate work.

A molecular machine can be constructed from any number of component molecules. However, compatibility between components of a molecular machine is decided by three parameters. These parameters are energy, sizes, symmetry or shape of individual components and complementary between them. These parameters also decide performance of a molecular machine. Properties related to molecular recognition needs attention to construct molecular machines from multiple components. In multi-component systems, if one interacting component is in a state where insignificant or no change in conformation is required to have efficient interaction with another component, it is referred to be in pre-organised state. Pre-organised state is a key factor to influence operation of multi-component machines. It relates free energy changes of a machine in terms of forward and backward action and also has implication in the entropy associated with the molecular motion in terms of free and bound state. Mechanism of interactions between two or more components generally passes through activation barriers. Activation barrier involves the conformational adjustments to provide a complementary state to bind to another component. The conformational adjustment process is energetically unfavourable; once conformational adjustments of a particular component take place, interactions of such a component with other component becomes energetically favorable. Thus, if energy associated with two components in reorganisation and interactions are comparable, then condition for reversibility can be achieved by application of internal or external stimuli. There are certain processes in which reorganised structures are created by stimuli to generate a reorganised state of a component conducive for interaction with another component. Simple illustration for such effect is conformational locks observed in molecules by weak interactions. Locked or an open conformational state can be equilibrated by applying external stimuli or by using another substrate which can interact with the parent component/s. For example, an amino acid has different conformers which differ in energy as illustrated in Figure 1.5. It has been observed that the presence of polar functional groups at the side chains of amino acids increases the number of low-energy conformers drastically. In naturally occurring amino acids numbers of such conformers vary, for example there are five such locked conformers for glutamic acid, six for cysteine and seven for threonine.

Description of molecular machines require two most important terms that are (a) Mechanical device and (b) Motor. Any device used for

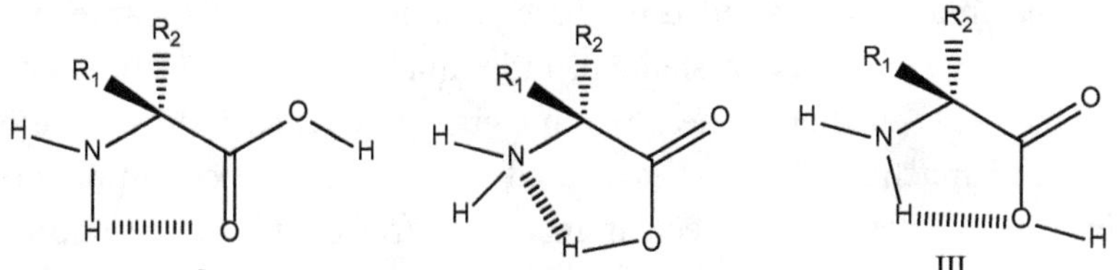

Figure 1.5. Some ways to lock conformation of an amino-acid (R_1 and R_2 are substituents).

mechanical movement is a mechanical device. Machines are subset of mechanical devices which utilises energy to perform a specific mechanical movement. On the other hand, a motor uses a given or available energy from surroundings to show output as work for specific and useful purpose. To generate a reversible operation turning on and off is a basic need of a machine; this important aspect is done by a switch. At molecular level a switch relates reversible modulation of properties in a multi-state system. In a molecular system reversible path from back and forth or circular paths can be related to switching process. A path is generally followed in large numbers of molecular devices to accomplish switching between multiple states, but such reversible switching process lacks feature; such an unique require additional work to be called as a machine. Hence, strictly saying, molecular systems showing reversible switching properties may not be called as motor but better to describe them under category of molecular switches. However, fundamental understanding of molecular switches goes along with as the fundamental concepts to develop a molecular machine, thus their understanding as an integral approach to molecular machine get priority.

1.5 Bio-inspired Concepts in the Development of Molecular Machines

Bio-inspired concept in miniaturization or making a gear comes from natural action of gears assisting in jumping action by certain insects.[10] Figure 1.6 illustrates the positions of joints of legs of a nymph at the time when it is about to jump. It is apparent in this figure that positions of adjoining portions of trochanter and femur change while it jumps. There

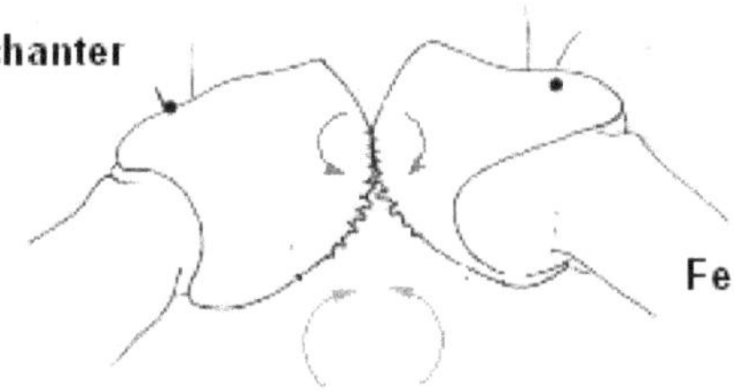

Torchanter

Femur

Figure 1.6.   Gear-like action of front leg joints of Lymph while jumping.

are two principal stages that help in such a jumping process. One stage is called preparatory stage and other stage is called propulsive stage. During preparatory stage both hind legs are rotated forward and hind legs are also elevated at their joints. In propulsive stage leg joints are depressed as shown by green arrows in Figure 1.6. Such changes in orientations take place at a rate with angular velocities 200,000° per second. Grid-like structures on the trochanter parts make adjustment to set up a gear-like action (red arrows in Figure 1.6). In both these stages, teeth like arrangements on bone which resembles parts of a gear helps to rapidly move in a synchronous manner.

Certain microorganisms named magnetotactic bacteria orient and migrate along lines of forces of the geomagnetic field. This unique feature is due to presence of membrane bound crystals of magnetic iron minerals. The iron minerals found in the magnetotactic bacteria are of nanometer-size.[11] These particles are found to adopt organized chain-like structures. The directional movement of magnetotactic bacteria has relevance to develop artificial materials to mimic magnetic properties shown by them. Such material would have ability to provide directions to a mechanical process as seen in movement of these bacteria. Movement of these bacteria is unidirectional towards north-pole of earth; hence, these classes of bacteria are named also as magneto sensitive bacteria. Inherent magnetic fields shown by these bacteria are associated with unpaired electron spin of paramagnetic substances; moving action of the bacteria is like performance of a living machine fueled by magnetic field of earth. Thus, from an analogy of biological systems, hypothetical crystal of material showing macroscopic gyroscopic property may be imagined at molecular level. It would be necessary to control molecular alignment of a crystal of a molecule to function as molecular gyroscope.

This is required as the directions of rotation of all molecules present in solid state are found as ensemble of molecules. One way for such an approach to align molecules is to utilize magnetic dipoles to improvise them as the rotary element of a molecular gyroscope. Under such a situation, a substance showing gyroscopic behavior is comparable to a compass. Bulk magnetic properties of such molecular compasses will be affected by an applied external electric, magnetic or electromagnetic field. Besides these, systems in solid state showing ferroelectric and anti-ferroelectric phases may also generate phonons to propagate at lower speed than sound.

New design, performance, type of fuel used and market economy are some factors which guides utility of machine in day to day life. A machine working with remote controls are of high demand for application. Such remote control machines are routinely used for all kinds of transport systems. Miniaturisation, integration and automation of machines are being universal issues that are required to make a better machine for utility as day to day item. Among these factors miniaturisation has several fundamental needs. First of which the scale at which the machine would provide maximum work with least expenses. In other word economy relating fuel consumption, waste management and ruggedness are factors that will have to account for in designing a new machine. A machine built with understanding of the performance of each component is like molecular understanding of ensemble of molecules performing a specific task in a united manner. Function of living beings to perform work is thus like an integrated machines. In a biological machine brain controls input and output. Volume or size-wise brain is a small portion of a body. This point itself provides impetus to miniaturize memory and signal transduction devices of a conventional machine so that a miniature component of a machine with highest priority controls or regulates inputs as well as outputs.

The action of living species, such as movement of a caterpillar on a sunny day can explain a locomotive machine. On a sunny day, a caterpillar take shelter in a place where shade is present but it comes out of the shade as the heat from sun gets lowered. When action of hot sun is less, which is in early hours of morning or at late afternoon hours, it takes a position in an open place. By such change in position it gets pleasant feeling of sunlight at those hours when sun is not strong. In this example, movement of a caterpillar is like translational movement of a body between two positions. Caterpillar is thus comparable to a machine, places covered while moving from shade to the bright spot is the transverse path and heat supplied by the sun is the stimuli.

All plants sense and respond to mechanical forces. Leaves of many plants are sensitive to touch or any other mechanical disturbances. A mechanical stimulus generated by touch on stem of certain herbs act as stimuli to generate mechanical motion of floral parts or leaves. Such stimuli or mechanical disturbances are also often produced by dynamic natural processes such as rain, wind, earthquake and gravitational force. These responses are used to self-protect or for acquisition of food by certain plants. Some of such responses are distinct and visible to naked eyes. Certain processes are also specific in direction. Actions may be taking place at a remote part irrespective of the direction from which a mechanical effect is applied. As an example, leafs of Mimosa pudica herb has open normal leafs under ordinary condition as shown in Figure 1.7a. On a rainy day or a highly windy day leafs are generally closed. This is due to sensitivity or response of leafs of this plant to mechanical disturbances. Alternatively, open leafs also closes when touched by human or animals or birds such plant (Figure 1.7b). Closing of leaves is independent of the

(a)   (b)

Figure 1.7.   (a) Leaves of a Mimosa pudica; (b) Same leaves when mechanically disturbed.

Figure 1.8. (a) An insect approaching an open leaf of a Venus' Flytrap. (b) Ejection of hairs in leaf once an insect approaches. (c) Trapping of an insect by folding a leaf.

direction of the stimuli where it applies, but depends on the strength of the stimuli and on the proximity of the leaf where stimuli have been applied.

Many plants close their stems, leaves or floral parts to trap insects. Venus Flytrap (botanical name Dionaea muscipula) is a carnivorous plant. This plant has bi-lobed leaves with needle-shaped tines on leaf margins. Leaves remain open and spread under normal condition. When an insect crawls along ventral surface of a leaf, it triggers hairs as well as it closes the leaf to trap the insect as shown in Figure 1.8.

1.6 Energetics and States of Molecular Machines

Machine works to show output by conversion or without conversion of one form of energy. Even the energy that is taking as input is not converted to another form but input stimuli is the guiding factor for an external stimulus guided machine. There are also machines where fuel is to be continuously given for consumption. Another set of machine may have source for internally generated stimuli or have a fuel source within. Hence utilization of fuel and equating such consumptions to energetic needs understanding of functional behavior of each component. But in a molecular machine such energetics has to be understood from molecular and electronic levels. The concept of building molecular machine starts with three major steps. First step is the synthetic aspect to make a designed unit. Second step is to understand properties associated with such a design. The later point clearly states workability of the machine for a particular purpose. Third step is to reason out the type of stimuli that will make it operational or to know if there is preference for a kind of energy required to run the molecular machine. In general a molecular machine is a device which produces useful work through understanding of electronic and structural aspects at different states that relates thermodynamic, kinetic stability. Hence there is utmost need to understand inter and intra-molecular interactions of each and every component molecules constituting the molecular machine. The intermolecular interactions can be between individual molecules or a set of molecular assembly or solvent-solute interactions. Equilibrations of inter- as well as intra-molecular interactions are widely utilized in operation of molecular machines. It is also well known fact that the properties studied at nano-dimension widely differ from the bulk materials. Due to such special properties associated with molecules or matter at nano-dimension this property explored at nano-dimension has distinct advantage. Moreover, different techniques have been developed to handle and study materials at nano-scale. Thus, nano-machines are machines in a molecular scale that suits one to study and improvise to get good output. Ionic and Van der Waal's forces, hydrogen bonds play pivotal role in designing a molecular machine. Geometrical changes of individual molecules or those of assemblies of molecules are successfully utilized to construct molecular machines. Ethane is a molecule where two methyl groups are held by a C-C bond. It has conformations such as Eclipse, Staggered forms (Figure 1.9). Energies associated with both forms are different and staggered form is more stable from the eclipsed form. Energy barrier can be crossed by providing required energy. Analogously, by lowering energy a particular conformation can be stabilized. Thus, the rotational motion of ethane makes each individual molecule like rotating a molecular machine. Another similar example is biphenyl, where C-C rotation leads to different atrop isomers. The activation barrier for such processes is guided by substituent effect and different steps can be equilibrated to construct a machine. Similar arguments are applicable to *cis-trans* isomers, syn-anti conformers, and optical isomerization etc. Under suitable conditions inter-conversion of each

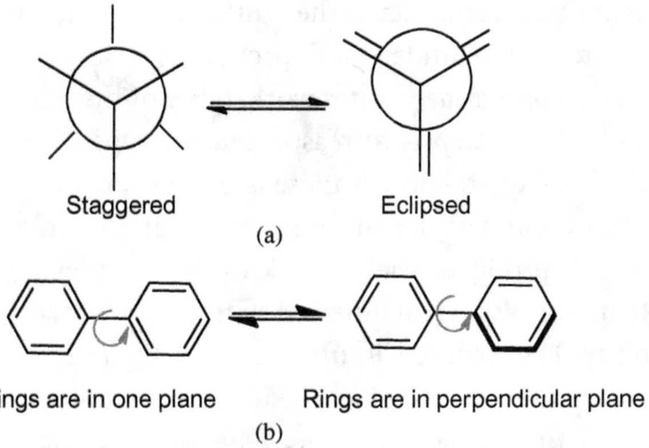

Staggered   Eclipsed

(a)

Rings are in one plane   Rings are in perpendicular plane

(b)

Figure 1.9.   (a) Staggered and eclipsed form of ethane. (b) Two conformers of biphenyl.

state are widely used to understand and illustrate principles behind function and design of a molecular machine. These examples on equilibrating properties have molecular understanding to control as well as to guide them for a particular function as per a need. Moreover, such equilibriums have certain characteristics and their utilities are predictable; thus they can be basis for designing molecular machines. Rotation of a C-C bond of methyl benzene (Toluene) is an example of a top moving over a fixed unit (Figure 1.10). Here the fixed unit is the phenyl ring and rotating top is the methyl group. Alternatively, the benzene ring has more numbers of carbon atoms relative to a methyl group hence benzene ring can be considered as pivot. Energetically, toluene molecule with one C-H bond of the methyl group lying in a plane perpendicular to the benzene ring is in a different state than the remaining two C-H bonds lying below the plane of benzene ring. These two C-H bonds have energy that is about 14.0 Kcal/mol lower than the energy of the other C-H bond that is perpendicular to the benzene.

Since activation barrier is small; equilibration of these two positions of C-H bonds are possible while methyl group rotates over the benzene ring to making each C-H bond indistinguishable under ordinary conditions. A similar situation can be extended to 1,4-dimethylbenzene (*p*-xylene) or other similar positional isomers. Extending such an idea one may imagine constructing a gear like arrangement from hexamethylbenzene[12]

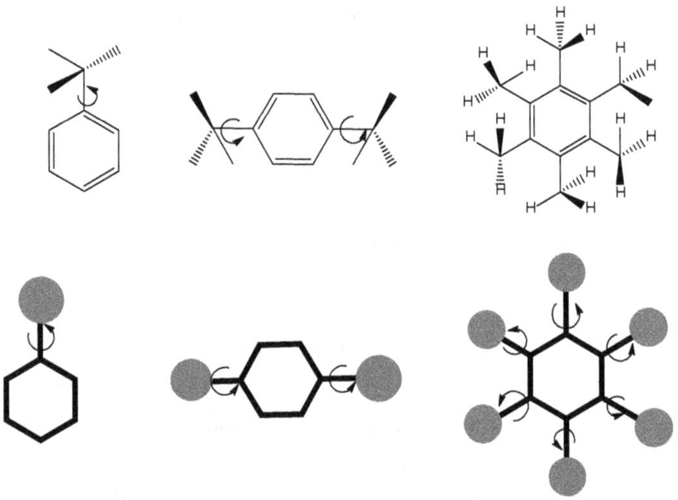

Figure 1.10. Models of top/s rotating over a fixed frame; (a) Toluene, (b) *p*-Xylene, (c) Hexamethylbenzene.

(Figure 1.10). It has six methyl groups, each undergoes rotational motion. But, to show property of a gear, it should show concerted directional rotations and one should be able to control such motions. The C-C rotation has activation barrier less than 10 Kcal/mole. This energy is comparable to solute-solvent interactions. Thus a solute-solvent interaction, namely, hexamethylbenzene interaction with solvent should be able to control C-C bond rotation of this compound. Rotations of six methyl groups are completely independent. Hence, there is a need to know about any correlated motion present in the molecule. A situation at which all methyl groups of hexamethylbenzene rotate in same direction is called co-rotating or anti-gearing. In another situation, methyl groups located next to each other rotating in opposite direction is described as counter-rotating or gearing mode. Gearing mode with six methyl groups performing together is categorised as hexagonal gear. Similar to toluene in the case of hexamethylbenzene also each methyl group occupies a position of highest potential energy with one C-H bond in the molecular plane, and one C-H bond above and one below the benzene plane. Due to such orientations there are four different energetically distinct states. As shown in the Figure 1.11, there is a stage in which all methyl groups are in same alignment with description of each methyl group shown as (a) in Figure 1.11. In second state, three similar alternative pairs of configuration as shown in Figure 1.11 (b), and third state is with two sets of methyl groups each having three molecules in the orientation shown in Figure 1.11 (c). Finally, fourth energy state is from six equivalent methyl groups organised in an orientation shown in Figure 1.11 (d).

As mentioned previously, to perform a particular function either as counter gearing or anti-gearing motion, energy barrier for these quantities must differ. Rotational energy barrier of gear motion must be higher than the anti-geared motion to show a gear like action by hexamethylbenzene. Theoretically, energy difference between two orientations shown in Figure 1.11(a) and Figure 1.11(b) is 0.4 Kcal/mol whereas the energy required to initiate gear-like motion is 33 Kcal/mol. On the other hand, gear slipping motion of a methyl group with respect to two adjacent methyl neighbours (Figure 1.11c) is 0.6 Kcal/mol. The theoretical calculation shows that concerted unidirectional rotation of neighbouring methyl groups in hexamethylbenzene is significantly lower than the motion of a methyl group along with adjacent methyl groups in counter-rotating manner. Thus, based on this data it is possible to obtain gear like activities of hexamethylbenzene by using a confined medium. Gear-like actions are shown in carbon nano-tubes.[13]

Self-assembly or multi-components interlocked molecules may perform work as a molecular machine. A molecular machine may be solid,

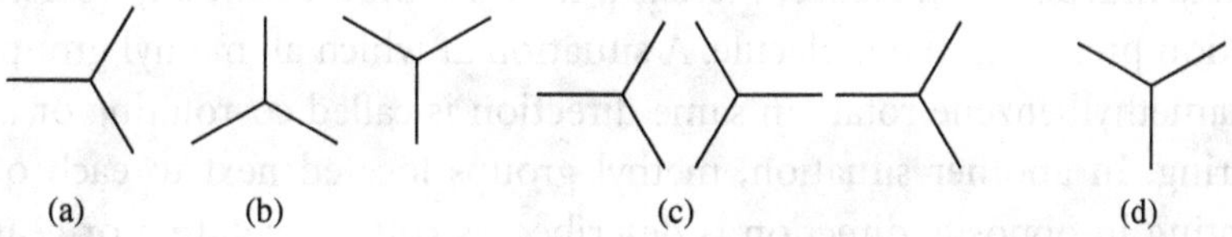

Figure 1.11. Different energy states from different orientations of (a) six similar orientations (none of C-H is perpendicular to the benzene), (b) three pairs with opposite orientations, (c) two pairs from combination of two similar orientations put in between by a different orientations, (d) six similar orientations (one C-H is perpendicular to aromatic ring).

liquid or gas. They may be also in other forms such as colloidal, micelle, liquid crystal, gel or solution. A molecular machine may perform work at an inter-phase between two states, as film, gel, gas, liquid or solid. Each of these has its own characteristic ways to show their performance. In order to perform a work in all these instances there must be changes arising from either from making and breaking bonds or from changes in scheme of original weak interactions, or from changes in equilibrium between phases or between different structures.

From experience and capability of performance of robots in real macroscopic world, one may imagine to miniaturize robots to a molecular level that would be called as molecular machine. Such a robot or machine will be performing at microscopic domain at molecular level, will have properties guided by electronic and architectural aspects of chemical bonds. They will be transformed to some other form by chemical reactions and leaves scope to modify or multiply them to do multiple functions. Making such molecular machines or devices from a unitary molecule or from molecular complex or from assembly as replica of robots provides many advantages to work at microscopic world. Such machines also serve as models to mimic biological functions. In the case of robot movements and performances are programmed, similarly in a molecular machine also defined activities of microscopic objects can be programmed. As mentioned it is convenient to fabricate properties of molecular machines at nano-dimension. Functional activities of many surface related properties are most efficient at nano-level; molecular machine performing at nano-dimension will be delivering in an efficient manner.

1.7 Biological Molecular Machines

In biological systems, amide bond is an important bond, and this bond is important as constituent bond of a molecular machine. Amide bond can be functionalized easily to other functional groups; and restricted rotation associated with carbon-nitrogen bond of an amide bond is utilized to modulate molecular properties. Restricted rotation around carbon-nitrogen bond of amide group takes place due to resonance structure. It provides various conformations and is functionalized and can be hydrolyzed to generate smaller units in selective part of constituent macromolecule.

Amide bonds in proteins are called peptides, they are responsible to control overall self-assembling properties of protein and also have roles to provide tertiary, secondary hydrogen bonded structures. Natural molecular machines based on protein backbone are often come across in biology. Proteins are responsible for performing many tasks such as carrier for molecules to an active site to catalyze reactions. Present days high speed computation facility and the macromolecular crystallography coupled with high resolution nuclear resonance techniques have enable one to precisely understand many of the protein based functional process that contribute to molecular machine.

In biology, the functional units which may be comparable to machines are made of nanometer-size molecules. These units operate at ambient temperature, pressure and volume in soft and chaotic environment produced by the weak intermolecular forces. So long a living unit is alive there is ceaseless and random molecular movements created by such machines. It may be noted that the gravitational forces and motion of inertia are negligible in such examples. The thermal energy in many cases is good enough to overcome the viscous forces resulting from intermolecular interactions among similar or dissimilar molecules such as solvent. Though an analogy on the construction of a frame of a molecular machine can be brought to molecular level but the operating mechanisms at the molecular level are much different from the machines operating at macroscopic world.[14]

Conversion of adenosine triphosphate (ATP) to adenosine diphosphate (ADP) is the major energy source in biological systems.[15-16] Adenosine part attached to triphosphate is like a handle which connects ATP to an enzyme. The terminal anhydride bond P-O-P bond is broken to form adenosine diphosphate (ADP) and phosphate (Pi) to use energy stored in ATP. In biology, it is essential to have constant amount of ATP to serve as sink of energy. Such a reservoir serves as a source for routinely produced energy from ATP to ADP conversion. ATP in biological system is formed by catalytic reactions. These catalytic reactions are caused by an enzyme called ATP synthase. This particular enzyme is present in intracellular membranes of animal mitochondria, plant chloroplasts and in bacteria. ATP formed in biological system is transported out of mitochondria and used for various biological functions. One such action is function of muscle. In ATP synthase there are two motors having distinct functions and structures. These two motors function together. One is called F_O-ATPase motor which is hydrophobic *trans*-membrane and another is globular F_1-ATPase motor. Based on the biological sources different names are given to F_1-ATPase. For example, mitochondrial motors are called mF_1, chloroplast motors are cF_1, Escherichia colimotors are termed EcF_1 and motor found in thermophilic bacterium are known as TF_1. The term F_1 unit comes from Fraction 1 and other part of the machine is called F_O motor. Structure of F_O part depends on organism from which it is obtained. The motor F_O (where O is subscript capital O) is a fraction which binds to oligomycin antibiotic. Oligomycin has capability to block the proton channel of F_O motor. Such blocking effect can eventually kill a bacteria. Further to these, regulatory actions of ATP synthase vary depending on the source of organism. F_O motor is embedded in the inner mitochondrial membrane in animals and it uses ion-motive forces based on ions such as proton or sodium ion for its function. Conventionally the proton-motive force is used as descriptor for performing work.

F_1 motor has a central protein 4.5 nm long stalk-like an axle, it is termed as γ-subunit (Figure 1.12). The stalk is surrounded by units called α- and β-subunits which are placed alternatively. They act like blades of a rotor. A view from the top suggests that α- and β-subunits possess circular arrangements. Among these two units, the β-subunits are responsible for performing catalysis whereas the α-subunits are regulatory for binding nucleotides. Three different types of structures are found in β-subunits, they are: open, loose, and tight binding site as shown in Figure 1.13. These sites are designated as O, L and T in figure and the depression areas shown over each blade represent the binding sites. Open and loose forms are catalytically inactive. The ADP and phosphate ions are weakly bound at loose binding site. In tight binding site ADP and phosphate are converted to ATP. This tight bound site undergoes conformational change to form an open state. In this process newly formed ATP is released. Such a conformational change in the β-subunits enforces rotation of γ–subunit. The γ–Subunit serves as link connecting the F_1-ATPase which initially acts as loosely bound state with the formation of ATP from ADP. Phosphate goes

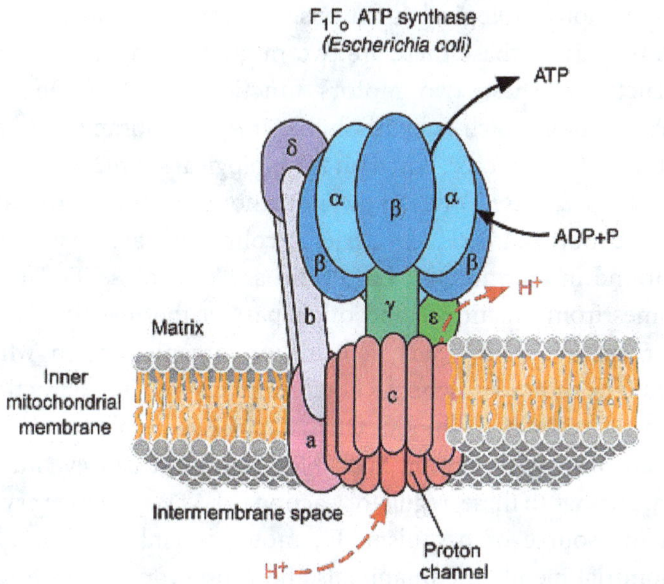

Figure 1.12. A schematic diagram of FoF$_1$ motor used in ATP synthesis.

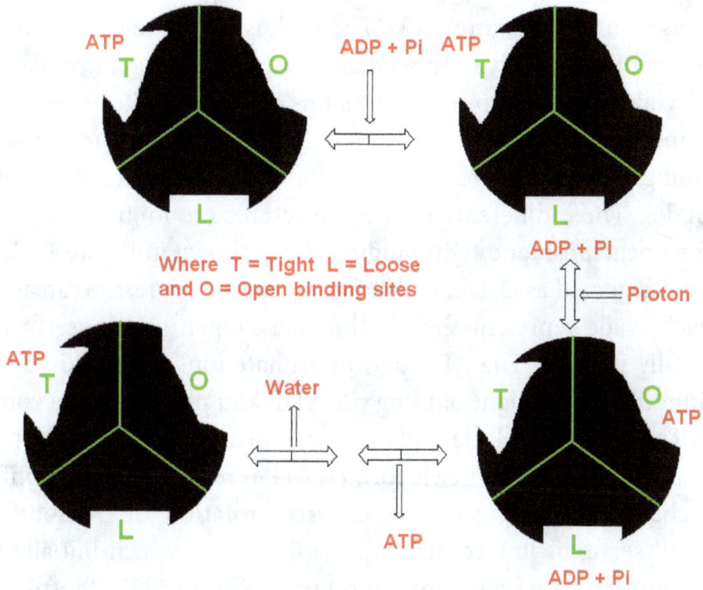

Figure 1.13. Schematic diagram showing the machine actions involved in ATP synthesis in biology by ATP synthase.

to a tightly bound state and finally open state is attained. F_1-ATPase can produce rotary torque of 80–100 pico-Newton nanometer. The rotation of γ-subunit plays a crucial role in formation of ATP.

F_O motor occupies space in the inner mitochondrial membrane of mitochondria. F_O motor is typically composed of a, b, and c subunits as shown in Figure 1.12. The portion a is a singular unit whereas b has two sub-units On the other hand, c has ten to fourteen subunits depending on source (Figure 1.12). While ions flow through membrane, it propels rotation of F_O-ATPase. During flow of protons a chemical potential gradient develops across the inner mitochondrial membrane. This potential gradient is responsible for function of ATPase enzyme to generate ATP.

Myosins, Kinesins, and Dyneins are three families of minute cellular machines in biology in addition to the ATP synthase.[17] Organelles, lipids, or proteins, are carried by these machines in biological systems. Conventional kinesin is a highly possessive motor; it is capable of taking several hundred steps without detaching. Helical two-stranded polymers of about 5 nm to 9 nm, comprised of actin protein are used as path for myosin-based molecular machines used as carrier. Such transport processes operate by consumptions of energy produced by hydrolysis of ATP. Myosines are also involved in generation of force to contract muscle. In muscle contraction thin actin filaments and thick myosin filaments slide. Kinesin and Dynein are proteins involved in biological cellular transport along microtubules. Myosin does biological cellular transport through actins. Microtubules of proteins that facilitate transport are about 25-nm diameter having opposite polarity at two ends. Movement of Kinesins and Dyneins in such tubules are in opposite direction, former move from the minus end to the plus end, whereas latter move from plus end to the minus terminal. Direction of tubules also varies, for example, nerve axons have such tubules in longitudinal direction with positive ends pointing away from the cell body; whereas in epithelial cells, positive ends point toward basement membrane. Dyneins exist in two forms: cytoplasmic and axonemal. Cytoplasmic Dyneins are involved in cargo movement, whereas axonemal Dyneins are involved in producing bending motions. The structure of Dyneins consists of two heavy chains in the form of globular heads, three intermediate chains, and four light intermediate chains.

Deoxyribonucleic acid (DNA) based molecular machine operating in nature are mainly used for information carrier. In DNA based machines there is no mechanical work as output. DNA structures are comprised of nucleotide bases, adenosine, thiamine, guanine, and cytosine adopting double helical structure. DNA molecules are applied in mechanochemical and in nano-electronic systems. A DNA double-helical molecule is about 2 nm in diameter and has 3.4 nm to 3.6 nm helical pitch irrespective of base pair and generally double-stranded DNA are about 50 nm length. On the other hand, flexible nature of a single-stranded DNA allows them to find application in machine components such as hinges or nano-actuators.[18] For utility purpose DNA generally finds advantage due to high ability for molecular recognition and to self-assemble. Due to complementary hydrogen bond formation as Adenosine-Thyamine, Guanine-Cytosine pairs, two complementary single strands of DNA in a solution recognizes each other to form double strand in specific manner. Free energy associated with complementary hydrogen bond formation in biological systems has been utilized to lift a hypothetical load.[19]

References

1. Browne WR, Feringa BL (2006) Making molecular machines work. *Nature Nanotechnology* **1**: 25–35.
2. Kay ER, Leigh DA, Zerbetto F (2007) Synthetic Molecular Motors and Mechanical Machines. *Angew. Chem. Int. Ed.* **46**: 72–191.
3. Erbas-Cakmak S, Leigh DA, McTernan CT, Nussbaumer AL (2015) Artificial molecular machines. *Chem. Rev.* **115**: 10081–10206.
4. Credi A, Silvi S, Venturi M (eds) (2014) Molecular machines and motors: Recent advances and perspectives *Topics in Current Chem.* Vol. **354**, Springer, Heidelberg.
5. Balzani V, Credi, A, Ferrer B, Silivi S, Artificial molecular motors and machines, Design principles and prototype systems (2005) *Topic in Current Chem.* **262**: 1–27.
6. Mavroidis C, Dubey A, Yarmush ML (2004) Molecular machines, *Annu. Rev. Biomed. Eng.* **6**: 363–395.
7. Balzani V, Credi A, Ferrer B, Silvi S, Venturi M (2005) Artificial molecular motors and machines: Design Principles and Prototype Systems. *Topics in Current Chem.* **262**: 1–27.
8. Hanggi P, Marchesoni F (2009) Artificial Brownian motors: Controlling transport on the nanoscale. *Rev. Modern Phys.* **E81**: 388–433.

9. Erbas-Cakmak S, Leigh DA, McTernan CT, Nussbaumer AL (2015) Artificial molecular machines. *Chem. Rev.* **115**: 10081–10206.
10. Burrows M, Sutton G (2013) Interacting gears synchronize propulsive leg movements in a jumping insect *Science* **341**: 1254–1256.
11. Yana L, Zhanga S, Chen P, Liu H, Yin H, Li H (2012)Magnetotactic bacteria, magnetosomes and their application. *Microbiological Res.* **167**: 507–519.
12. Burnell EE, de Lange CA, Meerts WL (2016) Molecular gears, *J. Chem. Phys.* **145**: 091101.
13. Han J, Globus A, Jaffe R, Deardorff G (1997) Molecular dynamics simulations of carbon nanotube-based gears. *Nanotechnology* **8**: 95–102.
14. Mallik R, Gross SP (2004) Molecular motors: strategies to get along. *Current Biology* **14**: R971–R982.
15. Boyer PD (1998) Energy, life, and ATP. *Angew. Chem. Int. Ed.* **37**: 2296–2307.
16. Walker JE (1998) ATP synthesis by rotary catalysis. *Angew. Chem. Int. Ed.* **37**: 2308–2319.
17. Roberts AJ, Kon K, Knight PJ, Sutoh K, Burgess SA (2013) Functions and mechanics of dynein motor proteins. *Nat Rev Mol Cell Biol.* **14**: 713–726.
18. Liu H, Liu D (2009) DNA nano machines and their functional evolution. *Chem. Commun.* 2625–2636.
19. Yin H, Wang MD, Svoboda K, Landick R, Block SM, Gelles J (1995) Transcription against an applied force. *Science* **270**: 1653–1657.

2

Operational Aspects
of Molecular Machines

2.1 Background

Legendary physicist Richard *Feynman* in a talk delivered at the American Physical Society in 1959 suggested that molecular machines could be possible by chemical reactions liberating energy when cold. Such a suggestion generated tremendous interest and inspired by biological systems have taken the study of molecular machine as frontier subject. Today, major portion of study on molecular machines are based on improvisation of fundamental reactions of organic and inorganic chemistry.[1-16] Ionic, covalent, dative and hydrogen bonds play important roles in performance of molecular machines. Performances and functional activities of molecular machines are in general controlled by mechanical, thermochemical, electrochemical and photochemical energy. Weak interactions relating to supramolecular chemistry become most important in understanding operational aspects of a molecular machine. Whether it is to build architecture of a molecular machine or study the inherent function of a molecular machine, each requires thorough understanding of supramolecular aspects. To perform mechanical works paths are required to be identified so that a transformation or equilibration between two or more energy states is possible within such paths. A path may be described by changing position of a body from one place to another to travel a distance, or it may be within a molecule or within an assembly without shifting the overall position from a fixed place.

Irrespective of dimension, every macroscopic or microscopic machine requires different components which serve supportive actions. For example, a motor requires just locomotive part but to do specific function such as to use the motor in night or to provide traffic signals lights are added. Light may be used as external energy but its intensity or wave length may be tuned by additional part of a molecular machine such as photo-sensitizer. Efficiency of such a machine may be changed by combining components that are photo-chemically or electrochemically driven. For safety one need to use a signalling unit like horn. This part also can be also an independent part or may be operated by the energy used in the motor. Similar concepts are true in biological or artificially designed molecular machines.

Certain basic thermodynamic principles related to machine, properties of media, state of matter of machine and machineries are needed to understand their functions. Functions of molecular machines are based on various equilibriums that are generally established with the aid of stimuli (heat, light and electricity). Progressive development of efficient machinery at small dimensions and improvements of experimental tools scientists have thought of converting force, motion, and energy within molecules, or in assemblies to redistribute within or dissipate to another molecule. Spectroscopic tools dealing with vibration, rotation, electronic and nuclear level have helped chemist to analyze and implement new ideas relating to molecular devices. On the other hand, translation and size at microscopic level could be realized by using optical equipment such as scanning electron microscope, transmission electron microscope, atomic force microscopes etc. Each of these non-evasive methods allows to precisely determining the movements and positions of any mechanized species. Furthermore, the excited state processes are no longer fantasy to scientists. The energy associated with excited states could be used to tune many optical properties related to material. Accuracy and deterministic domain has increased to provide precise data. Conventional mercury thermometers provide data on temperature at single digit after decimal. But, temperature measured by fluorescence probe is with much higher accuracy. In such cases change of temperature causing structural changes influencing highly sensitive fluorescence emission have been conveniently applied. For example complex **2.1** has an octahedral geometry, where two axial positions of the complex are occupied by two solvent (acetonitrile) molecules.[17] This complex in solution is blue when heated

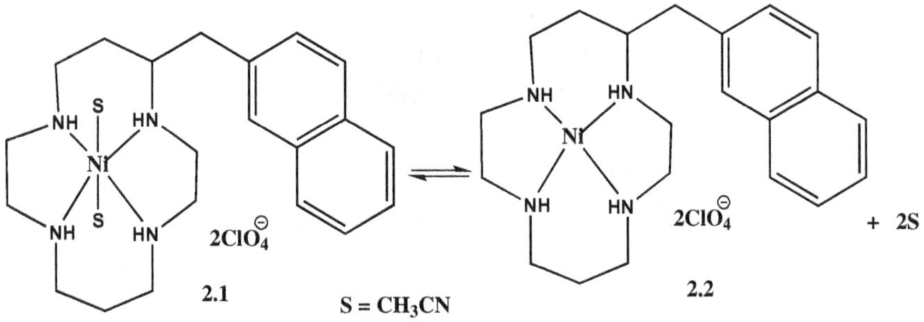

Figure 2.1. Modulation of fluorescence by temperature through reversible geometrical change.

it turns to yellow. This transformation is due to change of geometry to a square planar geometry around nickel. The complexes have nickel ions at +2 oxidation state (Fig 2.1). The octahedral complex is paramagnetic, whereas the square planar complex is diamagnetic. This nickel complex on heating in acetonitrile solution shows that fluorescence emission around 350 nm increases on heating in the range of 25°–75°C and on cooling it reverses the trend. Plot of intensity of fluorescence emission against temperature is used as thermometer with high accuracy. Spin-state of nickel ion affects quenching process. Similarly, changes in internal degrees of freedom of a molecular ensemble causing signal transductions are programmed to serve as molecular blueprint of a defined mechanical process.

2.2 Structural Equilibration

Supramolecular interactions are very important in devising a molecular machine. Changes brought about by performing a work changes original weak interactions scheme of ground state. Such changes in a molecular machine are reflected in various spectroscopic signals. Commonly, used tool to study characteristic feature in solution are nuclear magnetic resonance, Ultra-violet spectroscopy, Raman spectroscopy, Infra-red spectroscopy, Fluorescence spectroscopy and Electrochemical analysis. Micro-calorimetric titration is used a major tool for such study not only provides binding constants but also provides accurate information on various thermodynamic parameters such a entropy, enthalpy and free energy at microscopic scale. Due to tumbling motion in liquid, a property measured in liquid solution is independent of directions. Thus, measurements carried out in solution are time average isotropic property.

Any exchanging process generating signals from two independent states may be termed as slow or fast exchange. Life-time or rate of exchange between exchanging decides the shape of two signals at different temperature. Consider equilibrium between two different structures adopted by same compound and define the equilibrium as:

$$A \underset{kb}{\overset{ka}{\rightleftharpoons}} B.$$

Such exchange processes are routinely observed in NMR time scale. The equilibrium can be analyzed by recording NMR spectra and evaluated by approximating equally populated exchange sites. For an equilibrium between two states A and B, $P_A = P_B$, where P_A and P_B are populations of the respective state. The rate constant k for such an exchange process at a stage when coalescence of two signals occurs is directly proportional to the frequency difference of the two states. The coalescence temperature depends on system to system and is guided by activation energy of the process. The coalescence temperature is also influenced by solvent, applied magnetic field and nuclei under consideration. Let us say that A and B shows ^{1}HNMR signal at frequencies ν_A and ν_B respectively. Separation between these signals appearing at two frequencies $\Delta\nu$ is dependent on the life-time of the two species A and B. If the life-time of A is τ_A and that of B is τ_B, and one may define another quantity $\tau = \tau_A\tau_B / (\tau_A + \tau_B)$ where τ is a function of τ_A and τ_B; hence different types of signals for τ are observed under different situations. Namely, when τ is much greater than $1/\Delta\nu$ then two peaks will be observed and this process is called as slow exchange. When magnitude of τ is comparable to $1/\Delta\nu$, a broad signal will be seen, which is known as coalescence. Such a broad peak is centered at $(\nu_A + \nu_B)/2$. The different steps through which coalescence takes place belongs to a process called as intermediate exchange. On the other hand, when τ is much smaller than $1/\Delta\nu$ a sharp peak at $(\nu_A + \nu_B)/2$ is observed and this process is called fast exchange.

Rate constant and thermodynamic parameters are determined from such studies. Rate constant (k) at coalescence temperature of an intramolecular process is $k = 2.22\ \Delta\nu$. Incorporation of mutual spin-spin coupling (J_{AB}) between two interacting nuclei gives such a rate constant as $k = 2.22 \sqrt{(\Delta v^2 + 6J_{AB}^2)}$. Simulation of line shape analysis above and below coalescence temperature gives the activation energy associated with the process. When energy of A and B are different, then equilibrium constant (K) is defined either as the ratio of backward to forward rate or by ratio of population of the state B divided by the population of state A as follows:

$$K = k_b / k_a = P_B / P_A = \exp(-\Delta G/RT)$$

In such situation populations P_A and P_B are determined by using a weighted average of corresponding NMR parameters such as chemical shifts or J-couplings. The average parameter (π_{av}) may be calculated by using the an expression give below:

$$\pi_{av} = P_A \pi_A + P_B \pi_B = P_A \pi_A + (1 - P_A)\pi_B$$

where π_A and π_B are the weightage of the particular NMR parameter for corresponding form.

Consider equilibration between two chair forms of a six membered cyclic ketone 3-(1,3-dimethyl-4-oxopiperidin-3-yl)propanenitrile illustrated in Figure 2.2. In this example the average one bond C-C coupling constant can be calculated for a particular population ratio of A : B and the temperature dependent chemical shifts of such molecules help in studying equilibrium between the two chair forms.

Structural equilibrium may occur as a result of intra-molecular exchange of structures such as between tetrahedral with a square-planar(sp) structure, trigonal-bipyramid (tbp) with a square pyramid structure etc. Interchange

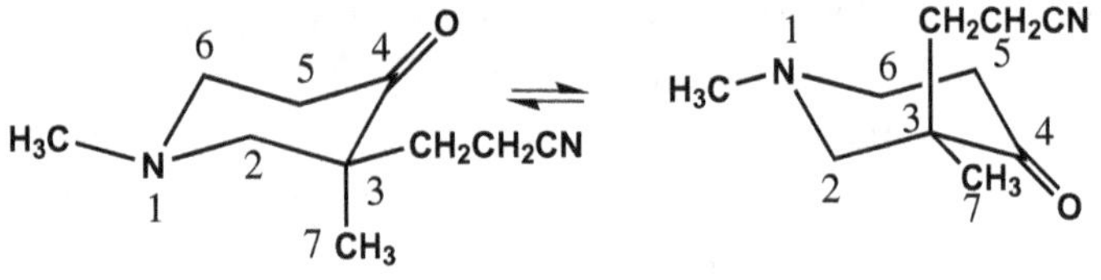

Figure 2.2.   Equilibrium between chair forms of a cyclohexanone derivative.

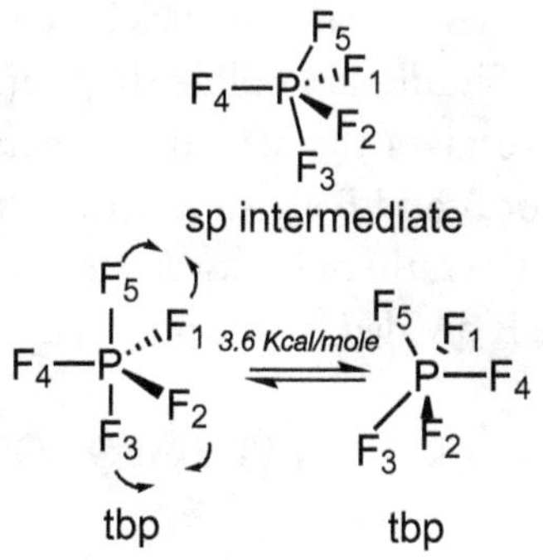

Figure 2.3. Phosphorus pentafluoride exchanging axial and equatorial positions (tbp = trigonal-bipyramid, sp = square-pyramid).

of axial and equatorial positions of tbp geometry also causes fluxional structures. PF_5 molecule is one such example which shows fluxional behavior. Activation barrier for inter-conversion of axial and equatorial positions of phosphorus pentafluoride as illustrated in Figure 2.3 is 3.6 kcal/mole.

There are two types of bonds in a trigonal bipyramid structure; one set comprising of three bonds are equatorial bonds which are on one plane, each adjacent bond makes 120° angle. There are also two other bonds perpendicular to trigonal plane and angle between these two bond is 180°. These bonds are called axial bonds. Transient intermediate step involved in exchange process of tbp geometry involves a square pyramidal intermediate. Such intermediate has four equivalent P-F bonds. Thus symmetry of the molecule changes from D_{3h} to C_{4v} during formation of transient species. Once this transient stage is reached angle between atoms that were originally axial decreases and angle between the atoms that were originally equatorial increases, until the axial atoms changes its position as equatorial positions and vice versa.

Tris-ethylenediamine cobalt (III) complex cation can be Δ or Λ optical isomer (Figure 2.4). Racemization of optical isomers of such tris-chelated complexes may pass through bond breaking or without bond-breaking. Mechanism of inter-conversion of two such optical isomers of tris-chelated metal without bond breaking mechanism has scope to equilibrate between two states. For such racemization process

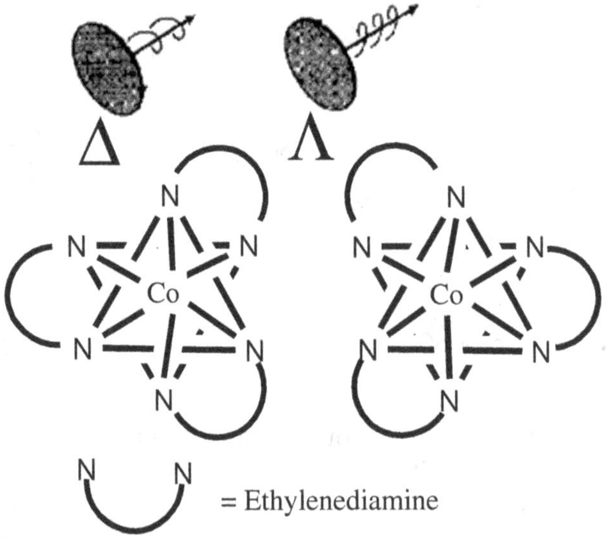

= Ethylenediamine

Figure 2.4. Optical isomers of tris-ethylenediamine cobalt(III) complex cation (inset is direction of spiral screw like path to locate ligating nitrogen atoms).

of tris-chelated octahedral complexes without breaking a bond passes through two conventional paths, they are named as Bailar twist and Ray and Dutt twist mechanisms. These two mechanisms namely, Bailar twist and Ray-Dutt twist mechanisms are independent of each other but common point is that each involves two consecutive $60°$ rotations of a triangular face of octahedral structure differing in the symmetry of intermediate species.

In solid state, $MoO_2(acac)_2$ (acac is acetylacetonate anion) exists as two chiral enantiomers,[18] namely Λ-*cis*-$MoO_2(acac)_2$ and Δ-*cis*-$MoO_2(acac)_2$. In solution, these two isomers (Figure 2.5) are in equilibrium and activation energy for such transformation between two forms is 68.1 kJmol^{-1} in benzene. ^{1}HNMR spectra at 27°C of $MoO_2(acac)_2$ has two methyl signals from the acetylacetonate ligand as singlets which appears at 1.58 ppm and 1.53 ppm. Variable temperature ^{1}HNMR spectra of a solution of $MoO_2(acac)_2$ in benzene-d^6 shows decrease in positions of chemical shifts (shielding) of the two methyl groups of $MoO_2(acac)_2$. Line broadening was also observed when spectra were recorded with incremental change temperature from 27°C to 67°C. Shifts observed in positions of chemical shifts as well as line broadening upon changing temperature are due to exchange

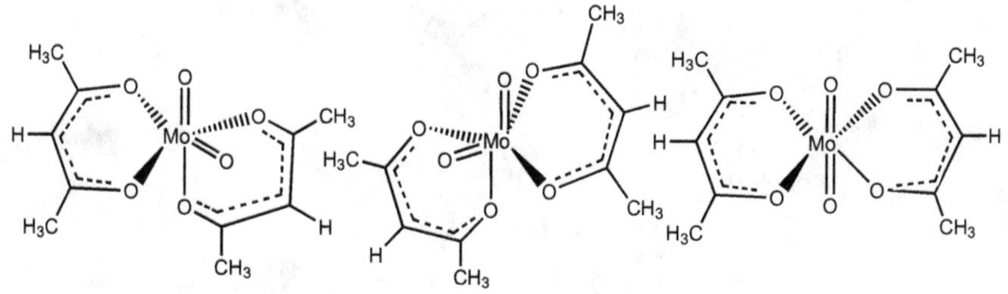

Figure 2.5. (a) Λ-*cis*-MoO₂(acac)₂, (b) Δ-*cis*- MoO₂(acac)₂, and (c) *trans*-MoO₂(acac)₂.

equilibrium involving positions of two methyl groups between two magnetically non-equivalent sites.

Mechanical rotor actions in molecules or assembly of molecules are generated through acid-base equilibrium. For example, 1,1′-Ferrocenedicarboxylic acid (**2.5a**) adopts locked structure through hydrogen bonds in neutral or acidic medium.[19] Whereas, in basic medium two carboxylic acid groups of ferrocenedicarboxylic acid transforms to anionic carboxylates groups (Figure 2.6). Hence, this deprotonation causes two carboxylate groups to move away to minimize repulsion between two anions. Iron ions present in between dicyclopentadienyl rings provide ball-like action to facilitate the movement. Another example of restricted C-N bond rotation is shown by taking an imide substituted quinoline derivative **2.6a**. When this compound is treated with an acid, it gets protonated and C-N bond rotation is hindered. In acid medium quinoline is protonated forming ⁺N-H bond. The ⁺N-H bond forms intra-molecular hydrogen bond involving Carbonyl group of imide, locks the conformer; which inhibits free rotation of C-N bond.[20]

Performance of molecular machine is greatly related to phase or state of matter of machine under consideration and also on medium in which it functions. But, most of the macromolecular machines are in solid phase working under aerial condition, or fluid and in conjunction of a fluid to increase efficiency and durability. A major portion of emphasis on study on molecular machine remains on their synthetic methodology, structural characterization. Synthesis and characterization aspects of molecular machine are conveniently studied in solution but lot of characterization are

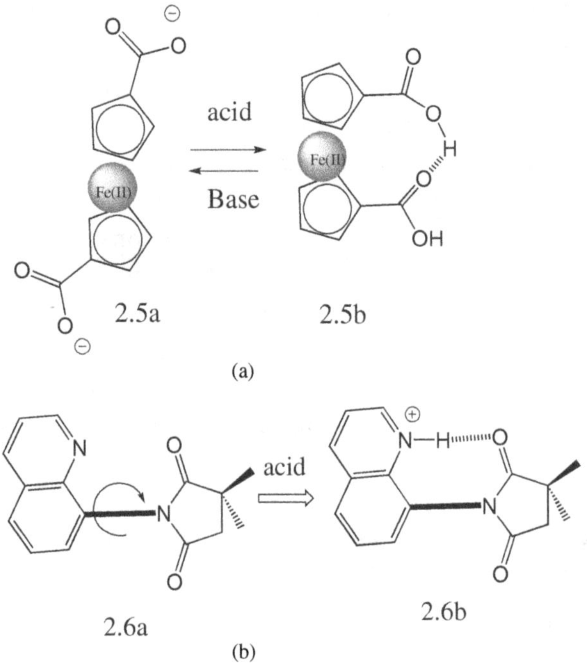

Figure 2.6. (a) Two orientations of ferrocenedicarboxylic acid at different pH. (b) Locking of C-N rotation by protonation.

done in solid which depends on state of the machine. Due to restricted motions in solid state, characteristic features of machine in solid state is very important. For diamagnetic organic based compounds or inorganic compounds multi-probe variable temperature solid state nuclear magnetic resonance technique is a very useful technique. Solid state measurements are also done with the aid of dielectric spectroscopy, and inelastic neutron scattering. Single crystal diffraction technique is one frequently used technique to characterize a molecular machine. Large numbers of chemically, photo-chemically, thermally executed reversible structural changes are studied easily by X-ray diffraction or by solid state NMR technique. Electronic properties of a component of a machine may change when it acts as part of a machine. So such changes require model studies to make a strong foundation for predictive properties. These unearth interesting aspects such as crystal to crystal transformation.

Topology related to various assemblies is a guiding factor to perform a particular operation by a machine. For example consider a channel-like structure or box-like arrangements with filled or empty spaces and the concept extends to carrier and delivery. There are other possibility that includes filled states can contribute as vessel or nano-reactor to provide new path for a particular reaction or to control equilibrium process within the voids. Such process may be also useful in construction of gates to navigate one substrate to another site and to control passage by opening and closing valves or walls. Thus, systematic endeavor to understand supramolecular structure and intermolecular dynamics contributing to physical properties of supramolecular assemblies provide guideline to construct useful machines. If a machine is characterized by solid state technique, it may be in crystalline form.

2.3 Molecular Machine Design

2.3.1 *Structural changes*

To design a molecular machine understanding of energetic of dynamic molecular systems as well as of functional supramolecular assemblies is a prime concern. For example, a rotating machine may be called as molecular rotor comprised of a molecule or part of a molecule that rotates against another part of the molecule. Such molecular motion may also occur on a surface with or without an anchoring unit called axle. When such molecular rotor is used to produce useful work; it is called a molecular motor. Rotors being one of the most important parts of a machine, they can be distinguished easily from the behaviour in medium. For example, a rotor may also operate in solution or vapour.

Octanitrocubane is an example of a rotor that operates in solution. In this molecule nitro groups occupies vertices of a cube (Figure 2.7). Orientations of nitro-groups permit this molecule to adopt different structures related by C4 or S4 axis. Activation barrier in each case is 2-5 kcal/mol[21] which is comparable to energy of solute-solvent interactions. Thus the modulation of the structures through crystallization is achieved.

Nitro group present in an inorganic metal complex also provide different orientations leading to different types of structures. Such observation

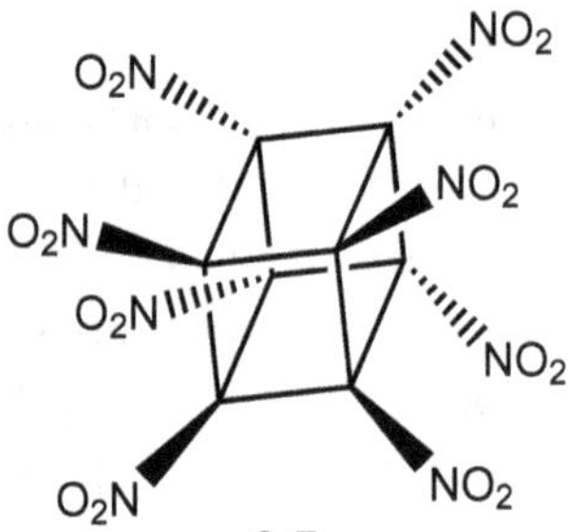

2.7

Figure 2.7. Octanitrocubane.

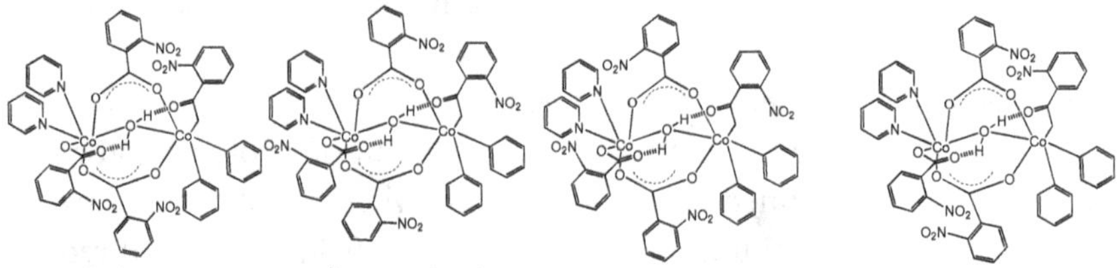

In-in out-out	In-out out-out	out-out out-out	In-in in-in
2.8a	2.8b	2.8c	2.8d

Figure 2.8. Projections of nitro-groups across two cobalt sites of a dinuclear aqua-bridged 2-nitrocarboxylate cobalt complex.

is found in dinuclear cobalt 2-nitrobenzoate complex.[22] Slight variation in reaction conditions in preparation of aqua bridge cobalt 2-nitrobenzoate complex results in complexes with different orientations of nitro-groups (Figure 2.8). Three polymorphic forms of such complexes characterised by single crystal structure determination are well known. Three isolated polymorphic forms have out-out, in-in, and out-in orientation of the nitrophenyl group with respect to dinuclear core. Interconversion of such forms can be done by crystallisation from different solvents.

2.3.2 Design of rotors

A rotor can be designed by placing rotating elements over a surface in several ways.[23] Simplest rotor comprise of three components, a movable part called rotator, an axle to hold movable part and a stand to anchor axle by providing a base. A stand is also called as stator that has much higher moment of inertia than the rotor part but a part of a molecule or an assembly of molecules connected to the axle (Figure 2.9). Simplest rotor having a rotating molecular unit (axle and rotator) connected to a surface is called as R-rotor. In a rotor where stator is rigidly connected to a surface to provide the necessary support to the rotor part, it is called as a RS rotor. Besides these rotational motion can differentiate a rotor. When rotator makes perpendicular turn with respect to the surface it is an azimuthal rotor or if it rotates parallel to surface it is an altitudinal rotor. However, for a rotor performing as crystalline solid, if the rotary components of molecules are part of the molecules it is a R-rotor, or if it has separated adjacent rotators it is a RS-rotor. Plastic crystals are R rotors, as in this state of matter solids themselves can perform rotary motion.

Although an axle is projected as a component of a rotor, a rotor may be also without a defined axle. For example, when a plate rotates over a static surface or in vacuum under influence of a field, it may not require an axle to rotate. In such a situation the axis of rotation may be imagined as virtual axle in such cases. Similar facts are also true in the case of inclusion complexes, in which adsorption and desorption may cause rotary motion of a guest molecule. In case of adsorption and desorptions the guest molecules do not necessarily require an axle as stator.

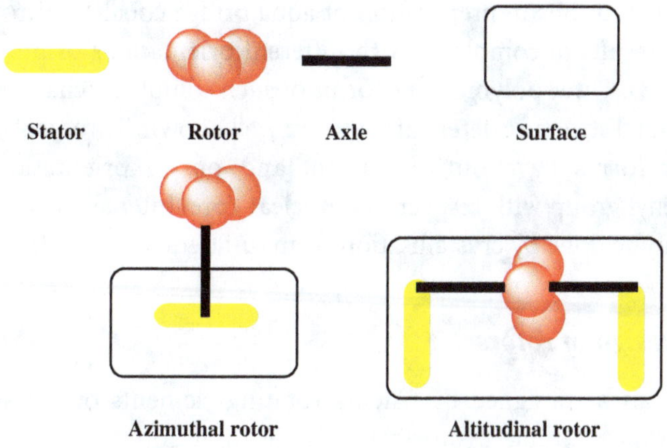

| Stator | Rotor | Axle | Surface |

Azimuthal rotor   Altitudinal rotor

Figure 2.9. Two types of rotors and components.

2.3.3 *Design based on conformations*

The cobaltocene complex **2.9** is an example of an azimuthal rotor. In this case the two legs are from piano-stool type cobalt-cyclopentadinyl parts holding a rotatable rod-like structure constructed from combinations of aromatic rings.[24]

The silicon derivative shown in Figure 2.10 has a rotatable arm for C-C rotation and has a stator part based on the tetraphenylsilane derivative

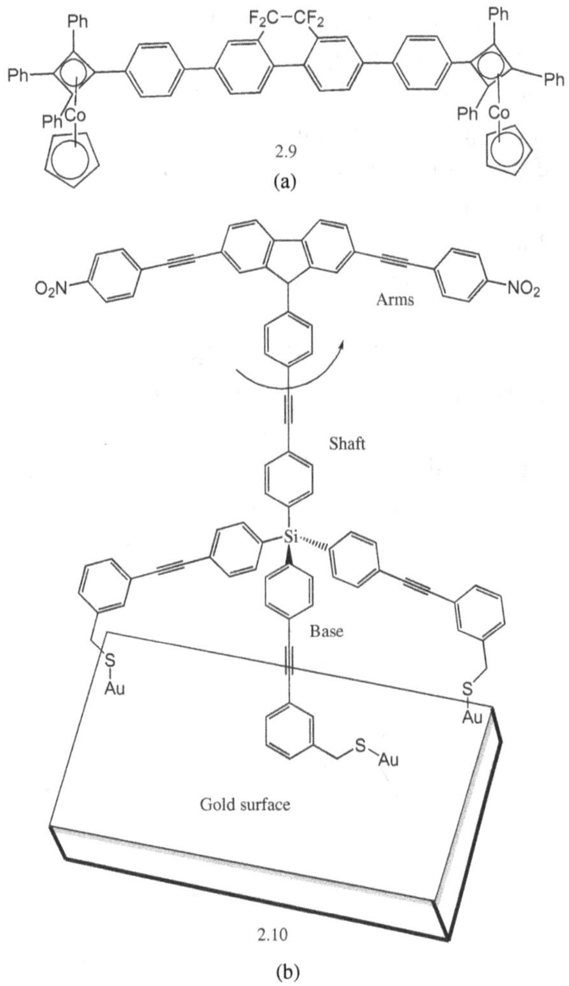

Figure 2.10. An azimuthal molecular rotor (**2.9**). (b) A molecular rotor with a stator (**2.10**).

which acts as tripodal base.[25] The molecule is fixed on a gold surface by three sulphur atoms through Au-S bonds. The molecule thus project upward from the modified surface of gold. The rotating arms of the molecule have a strong net dipole a required property to cause movement by electrical field. Pre-selection based on diverse combinations of such molecules are interesting for supramolecular rotors of macroscopic rotary devices for specific purposes.

Calix-4-arene shown in Figure 2.11 comprises of four 4-tertiarybutyl phenol ethers connected to each other at their respective ortho-positions through intervening -CH$_2$- group. These compounds are very selective to respond to ions and polarity of solvent and may adopt any of the three forms shown in Figure 2.11. In partial cone structure three ether units project in one direction and one on the opposite direction of the rim. Whereas, in the case of other two forms the rings are in opposite direction of the rim 1,2 alternating or in 1,3 alternating ring structures have the phenyl rings projecting ends in opposite sides of the rings are in pairs or at 1,3 alternate positions respectively. The advantage of such conformation may be taken to make cyclic rotating units fixed within a framework. Based on such principle a molecular mill is illustrated in Figure 2.12. In this example two independent calix[4]-arenes are connected by bis-crown ethers having intervening ether linked 2,3-hydroxynaphthalene units. Advantage of 1,2-alternate geometry of calix[4]-arene is taken as this form provides two hydrophilic end at 1,3-alternate position to functionalize

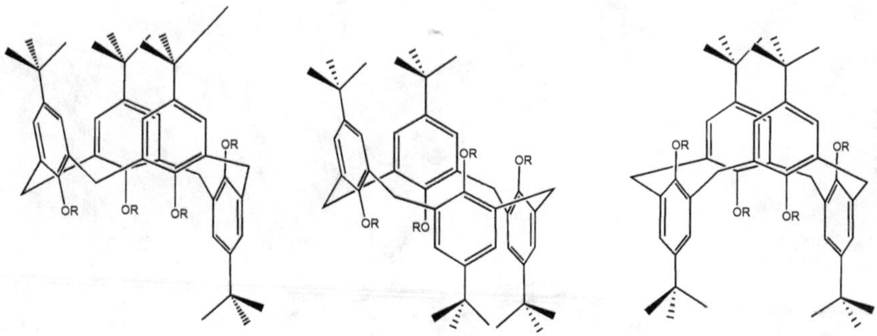

partial cone

2.11a

1,2-alternate

2.11b

1,3-alternate

2.11c

Figure 2.11. Three different forms of acalix-4-arene.

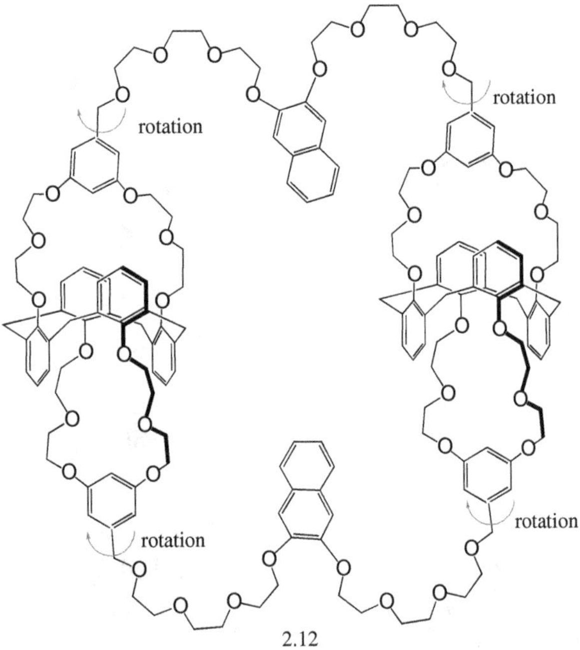

2.12

Figure 2.12. A molecule having architecture of a molecular mill.

them to lock the conformer. Molecular modeling studies on the system favors carbon-carbon bond rotation of calix[4]arene units as illustrated by arrows in the Figure 2.12. Such possibility in the molecule makes it a potential skeleton of a molecular mill.[26]

2.3.3.1 *Molecular brake*

Certain molecules have inherent shape and functional aspect to act like a component of a molecular machine. For example propeller shaped structure is formed when two bulky aryl rings are connected to one central atom with sp3 hybridisation (Figure 2.13). Central atom in such case is called as focal atom. Such molecules have rotatable axes connecting the focal atom.[27] A clock-wise rotation of one ring in such molecules induces a counter-clockwise rotation on the opposite ring; thus they behave like gear. The compound shown as 2.15 has a 9-triptycyl unit connected to a

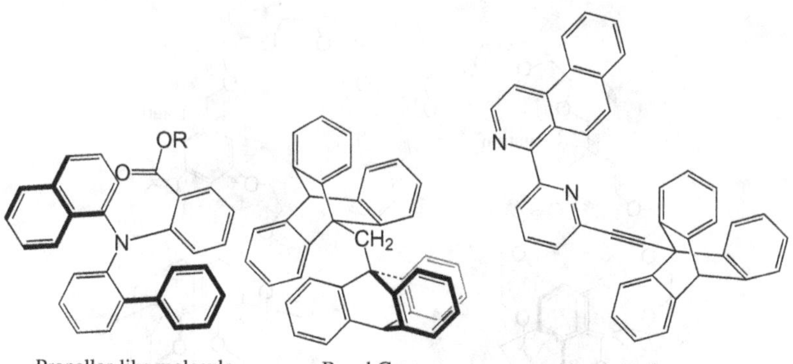

Propeller-like molecule	Bevel Gear	Brake
2.13	2.14	2.15

Figure 2.13. Some molecules having inherent shape of component of machine.

2,2′-bipyridine derivative through an acetylene group. The movement of 2,2′-bipyridine part of this compound is controlled by complexation and decomplexation with metal ions. Free rotations along single bonds are also controlled by changing orientations of 9-triptycyl unit making it to execute work like a molecular brake.

A molecular brake is easily constructed from a 2,2′-bipyridine derivative by taking advantage of C-C bond rotation between two pyridine rings. A rotation around the C-C bond provides either syn or anti form. Syn-form is the one which is present in metal chelate whereas anti-form is more thermodynamically stable in free state. Mercury complex of ligand **2.17a** resembles functional behavior of a brake. Syn-form of **2.17a** gets locked by chelation to the ligand through two nitrogen atoms of bipyridine rings to mercury ion (Figure 2.14). When ethylenediamine tetra-acetic acid (EDTA) is added to complex **2.17b**, it releases the ligand. This is due to formation of EDTA complex of mercury and once free biprydine ligand is released it adopts anti-conformation as that of **2.17a** by undergoing a 180° C-C bond rotation.[28]

Conformational changes in 2,2′-bipyridine is utilised to show three different states differing in dihedral angles between pyridine rings (Figure 2.15). For example, compound **2.18a** adopts a twisted geometry in free-state. Dihedral angle between two pyridine ring changes on complexation to potassium ion to form **2.18**; in this case nitrogen atoms are away from each other and planarity between two such rings are absent.[29] Whereas, nitrogen

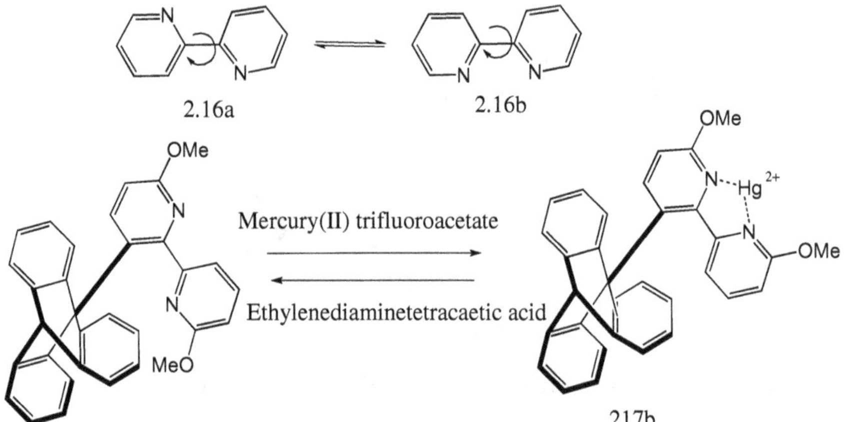

2.16a 2.16b

2.17a 217b

Figure 2.14. Two forms of 2,2′-bipyridine (Top). Bottom is example of a molecular brake.

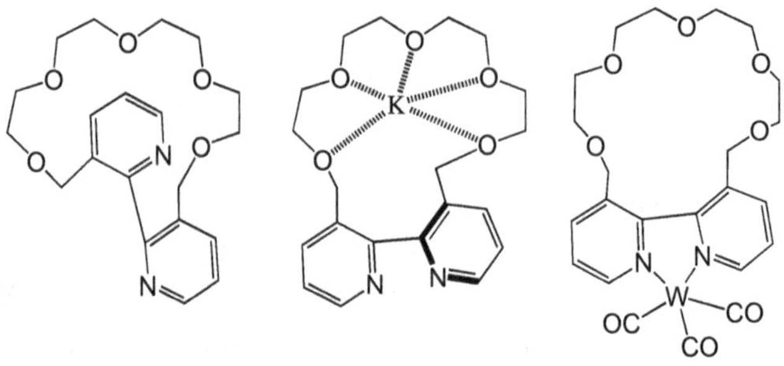

2.18a 2.18b 2.18c

Figure 2.15. (a) Crown connected to 2,2′-bipyridine, (b) Potassium ion encapsulated crown, (c) Tungsten carbonyl complex of the ligand.

atoms can be brought to project in same side and planar geometry between two rings is established by anchoring the bipyridine to a metal ion. In this example, complexation with tungsten tricarbonyl form **2.18c** to execute the conformational changes.

2.3.3.2 *Molecular bevel gear*

Functional behavior of bevel gear can be easily seen in the compound bis(4-methyl-9-triptycyl)difluorosilane[30] shown in Figure 2.16. Compounds **2.19a-2.19b** exist in optically active forms which are mirror image of each

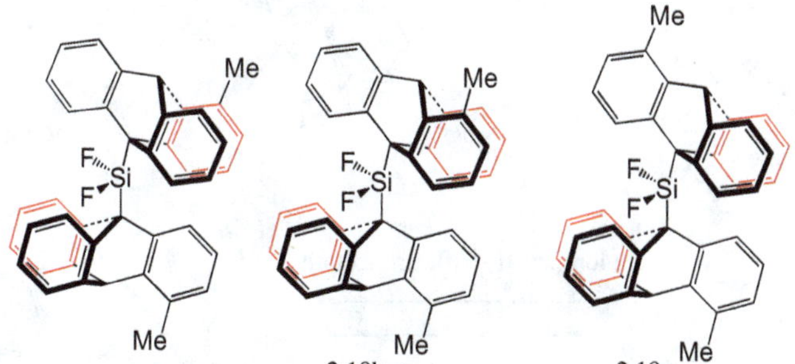

2.19a 2.19b 2.19c
Optical isomers Meso isomer

Figure 2.16. Different isomers of bis(4-methyl-9-triptycyl)difluorosilane.

other and it also exists as optically inactive *meso* form **2.19c**. This molecule function as gear system of a molecular machine with external controllable clutch-declutch. Use of fluoride as external stimuli to cause reversible equilibrium is in operation of such a gear is shown in Figure 2.16. Silicon atom adopts four, five and six coordination geometry in the curse of the process. Moreover, silicon has high affinity to form Si-F bond. Thus bis(4-methyl-9-triptycyl)difluorosilane reacts with supramolecular complex of crown ether encapsulated potassium fluoride to form a penta-coordinated species under non-aqueous condition. But when water is added fluoride ion is released from penta-coordinated species, this is shown in the right side of Figure 2.17. Once a fluoride ion is dislodged from the penta-coordinated complex original tetra-coordinate difluoro compound is formed. Thus, geometry across the C-Si-C axis chiral difluoro compound **2.20a** or **2.20b** is altered to provide a different geometry through expansion of coordination sites by the externally added fluoride ion. Such change is a prerequisite condition in function of a bevel gear. Hence, fluoride ion assisted change in this example is a functional model of a molecular bevel gear. A hydrolysis reaction of penta-coordinated compound **2.20c** causes rotational motion, which is another aspect that makes the equilibrium to complete all functional behavior associated with a bevel gear. Entire process involves bending, straightening and rotational motions.

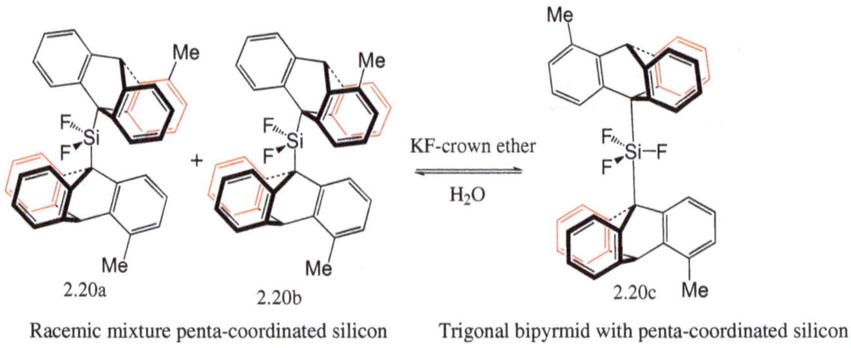

Figure 2.17. Rotation and bending to function as molecular bevel gear.

2.3.3.3 *Molecular rotating wheel*

An example of a rotating wheel based on a trypticene group is shown in Figure 2.18. Compound **2.21a** has a bend pawl and free rotation of trypticene part is hindered by steric reasons. The compound **2.21a** is easily functionalized to make it an amino derivative. This amino derivative **2.21a** possess a functionalisable hydroxy group. This compound adopts a helical structure and belongs to a category of compound called helicene. Hydroxy group containing part acting as pawl is suitable for unidirectional rotation. To cause unidirectional rotation the molecule is treated with phosgene so that it provides a reaction sequence that causes a control rotation as illustrated in sequences of processes shown in Figure 2.18. Entire rotation process[31] is described by combination of following processes: (a) conversion of amino group to isocyanate group by reaction with phosgene gas to form a compound **2.21b**. As this step facilitates the rest of the steps to cause mechanical motion, hence, this step is fuelling step. (b) In next step conformational adjustment by rotation on isocyanate compound takes place. This step makes the system ready (**2.21c**) for further reactions. (c) Third step is formation urethane (**2.21d**) by reaction of the hydroxy group with isocyanate group. (d) In the fourth step an irreversible rotation to cross the energy barrier to attain a suitable conformation **2.21e** takes place. (e) Lastly, hydrolysis of urethane **2.21e** regenerates original amine compound **2.21a**. Conversion of amine to isocyanate is followed

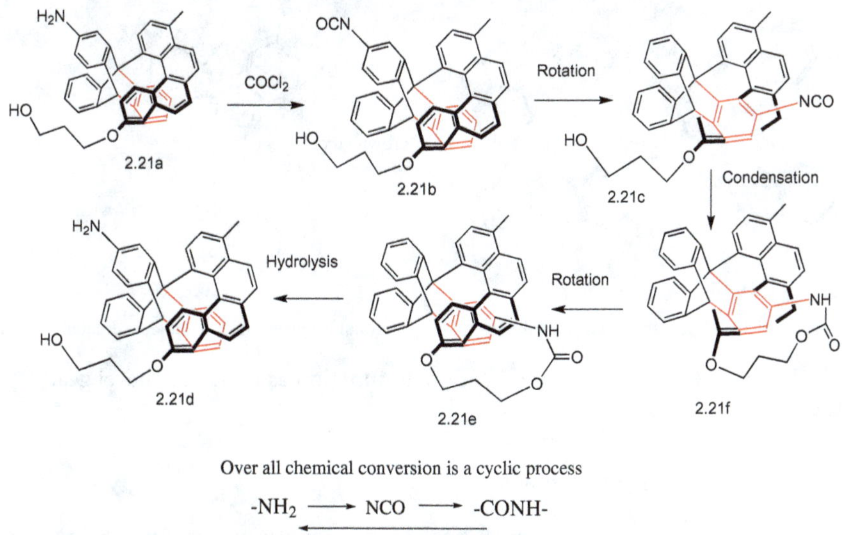

Over all chemical conversion is a cyclic process

$$-NH_2 \longrightarrow NCO \longrightarrow -CONH-$$

Figure 2.18. Functional model of a rotating wheel based on a trypticene, showing an overall cyclic process.

by formation of urethane and subsequent hydrolysis of urethane are chemical reactions where fuel is phosgene gas. These reactions produce hydrochloric acid as waste. Due to random movement of the flexible arms reactive moiety come into the proximity of the hydroxy group for urethane formation. As a result a fast intramolecular reaction to lock conformer and attain thermodynamically stable form takes place. Hydrolysis reaction disrupts the CO-NH bond to provide the unidirectional rotation over a C-C bond.

2.3.3.4 *Molecular tweezers*

Inclusion of a guest molecule in tweezers-like structure of a receptor molecule such as **2.23a** shown in Figure 2.19, can be stabilised by guest binding. The receptor has two anthracene units connected through rotatable C-C bond are bought parallel to each other by forming host guest complex with an electron deficient planar host. This host is accommodated between the anthracene units of the host so that a tweezers-like action is performed.[32] The host-guest complex when treated with one equivalent of copper(II) salt with respect to host, one bipyridine unit

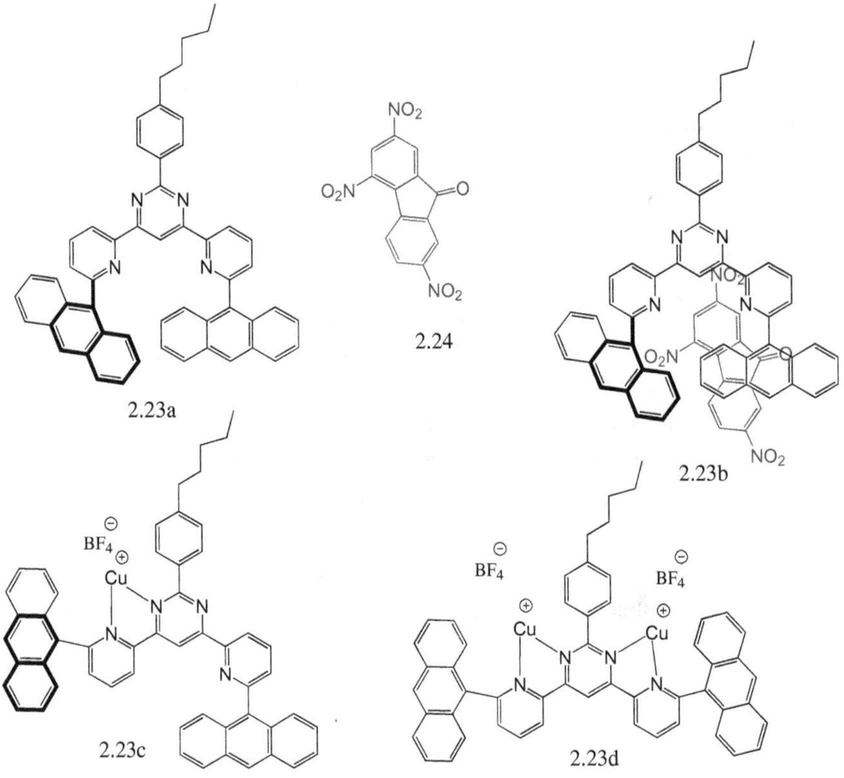

Figure 2.19. Tweezers-like orientations formed by molecular recognition and flattening by interaction with metal ions.

changes orientation to hold copper ion forming a chelated complex. Thus, anthracene units are no longer available to act as tweezers. One side of the receptor has different orientations of two freely rotatable anthracene units. However, same compound when treated with two equivalents of copper ions two anthracene units present as two arms of the molecule reorganises to adopt a flattened structure.

2.3.3.5 Rotors

Generally, a single crystal comprising of one component or multiple components is rigid, homogeneous and periodic and has macroscopic anisotropy. Tight packed structures are found in crystals due to presence of long range interactions in solid state of matter. Due compactness such structures have limited scope for lattice vibration and are content with small-amplitude lattice vibrations which is not enough to cause molecular displacements. There is also another class of crystals, which are called as plastic crystals. Such crystals are highly symmetric comprising of spherical or cylindrical shaped assemblies. In plastic crystal center of mass remains fixed on a given lattice position, and entire molecule experiences very fast rotational motion. These molecular motions occur either continuously or by jumping between equivalent sites. Adamantane, neopentane and many long chain alkanes are examples of plastic crystals.

Glass phase formation is another state of matter shown by many materials. In glass state molecules are highly inhomogeneous and possess many molecular conformations and intermolecular arrangements. Despite of these facts, physical properties of macroscopic glasses are isotropic. Liquid crystal is another state of matter where there is a decrease in molecular order. Liquid crystals have high translational, rotational and conformational freedom to changes by stimuli. Several other types of phases are observed in liquid crystals such as nematic, smectic, discotic phases depending on the type of molecules with rigid, rod- and disk-shaped cores. In liquid crystalline phase constituent molecules retain preferred molecular orientations and maintain average long-range molecular order. Molecules in liquid crystalline state also undergo fast rotation and diffusion. Thus, dynamics associated with these states have perquisite quality to support rapid motion of well-designed components. Hence, from overall understanding on different states of matter extending from domain of crystals to liquid, presence of rigid ordered components another state possessing highly mobile component state is defined as "amphidynamic" solids. The term amphidynamic originates from the word "amphi" which means "both sides".

Molecules having dynamic structures where rotary motion dominates is comprised of single mobile element or multiple mobile elements are individually defined as molecular rotor. Rotor part of such molecule has distinct characteristics of associated movement with respect to rest of the molecule. Such rotations are performed or occur with the aid of relatively small energy and energy used is generally comparable to thermal energy. In other words, a rotary action in molecule leading to orderly rotation requires a design principle. Based on design and function of a rotor, fine-tuning of properties of a molecular rotor is possible by variation of substituents.

Palladium complex shown in Figure 2.20 has N_2Cl_2 coordination environment around square planar palladium ion resembles a two-armed molecular gyroscope. This complex has a paddle like Cl-Pd-Cl unit, which is held in a semi-rigid structure.[33] The geometry of the molecule is such that a free rotation across the N-Pd-N bond axis is possible. On the other hand, another complex has a bulky penta-fluorobenzene ring. Bulky ring cannot easily pass through the void of the macrocyclic ether. Thus, this molecule shows back and forth motion between 180° positions. Substituted pyridine or symmetric disubstituted phenyl molecule as rotor part linked to rigid triple bonds connected to triphenyl methane at two ends is an example of a molecular rotor (Figure 2.21). Such

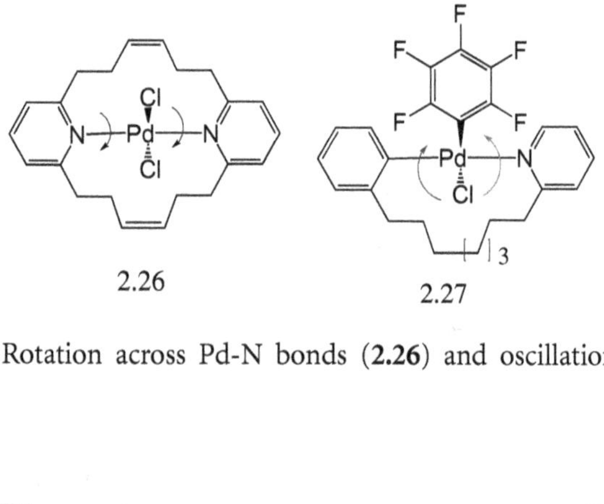

2.26   2.27

Figure 2.20.   Rotation across Pd-N bonds (**2.26**) and oscillation across Pd-N bonds (**2.27**).

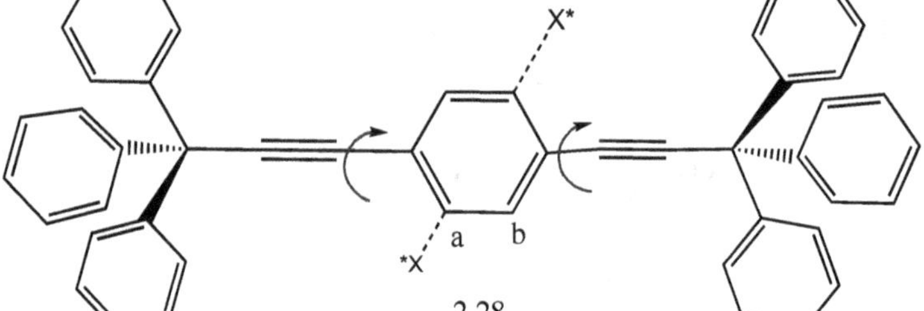

2.28

Probability of X and X* are equal

Figure 2.21.   Rotation making two C-X bonds equivalent.

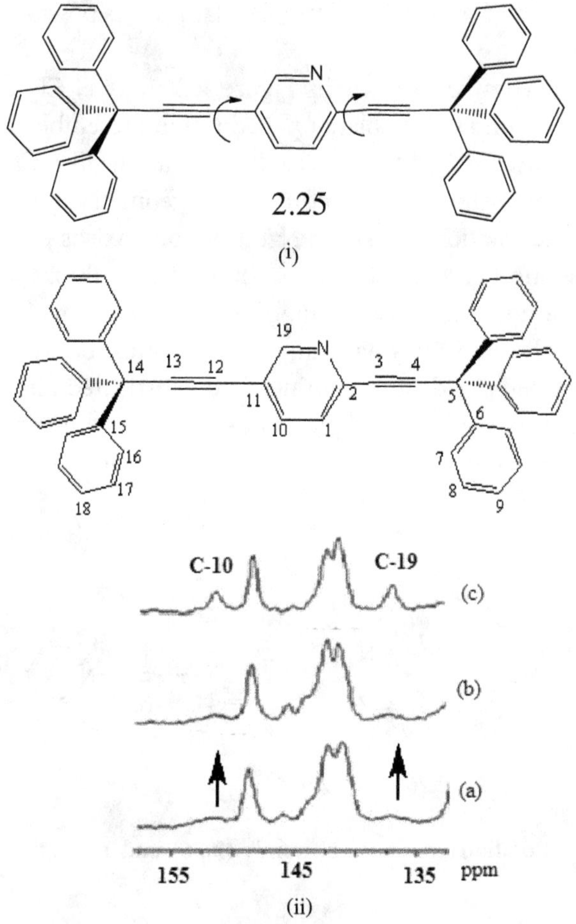

Figure 2.22. (i) Example of an amphidynamic compound; (ii) A part of solid-state ^{13}C-CPMAS NMR spectra in the range of 135–155 ppm of the compound recorded at (a) 113°C (b) 22°C and (c) −108°C.

molecules in crystal lattice adopt structural arrays in lattice like molecular compasses. Compounds **2.28** showed in Figure 2.21 in solid state shows temperature dependent changes in position of chemical shifts of the ^{13}CNMR signals arising from carbons of C-X bonds on the central rotating phenyl unit. This change is related to the rotation of the central ring. ^{13}CNMR signals at −118°C for carbon atoms labeled as a and b in Figure 2.22[4] are broad and as the temperature is increased to 113°C sharp peaks for these carbons are observed.

Compound **2.25** has a pyridine ring hinged to two acetylene derivatives providing scope to C-C rotation[34] as illustrated in Figure 2.21. Rotation of pyridine ring of this molecule in solid state is reflected in solid state ^{13}CNMR spectra recorded at different temperatures. As shown in the Figure 2.20 the signal for C10 and C19 are observed at low temperature, whereas such peaks broaden at high temperature and are not observed.

Structures determined by X-ray crystallography of such compounds that has rotatable aromatic ring shows crystallographic disorder. Such disorder occurs to show two equivalent occupancies with half the electron densities shared between two equivalent positions related by inversion symmetry. Thus, a C-X bond is observed by 2-fold flipping of the ring of an unsymmetrical aromatic unit. In Figure 2.21 C-X bond is shown by dashed line that has equal probability of finding at two equivalent positions due to rotation. Interesting aspects of this class of compound is that these molecules can be synthesized easily by conventional reactions of organic chemistry. A reaction scheme used for synthesis is shown in Figure 2.23. In the first step a triphenyl substituted acetylene is prepared by Grignard reaction. In the following step a catalytic cross coupling reaction between the acetylene with 2,5-dibromopyridine is done. The product of such a reaction on further treatment with meta-chloro perbenzoic acid gives the N-oxide derivative, which has also characteristic features of a molecular rotor.

Rotors are also constructed within cage-like containers;[35] one such example is shown in Figure 2.24. Dynamic nuclear magnetic resonance studies of the compound **2.31** supports rotational motion of the difluorophenyl unit as rotor. In this rotor a cage is formed by connecting flexible tethers at three ends of three independent phenyl rings connected to a central carbon which holds the rotatable difluorobenzene part by intervening acetylinic bonds. The difluorobenzene unit held to acetylene units through *sp*-hybridization can adopt different orientations within the cage.

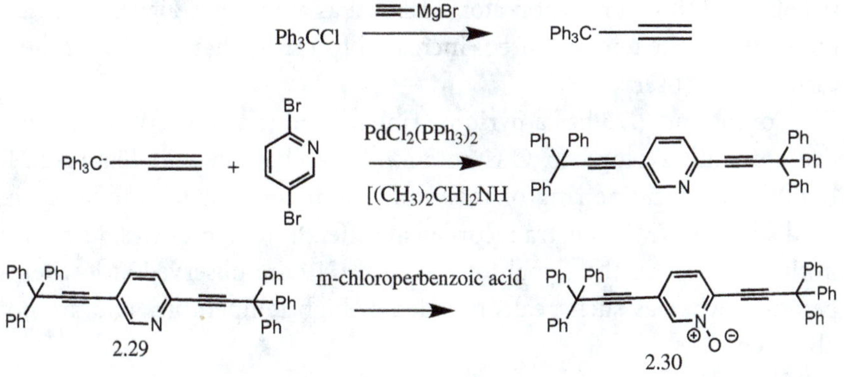

Figure 2.23. Synthesis of a molecular rotor having rotatable pyridine N-oxide flanked by substituted acetylene units.

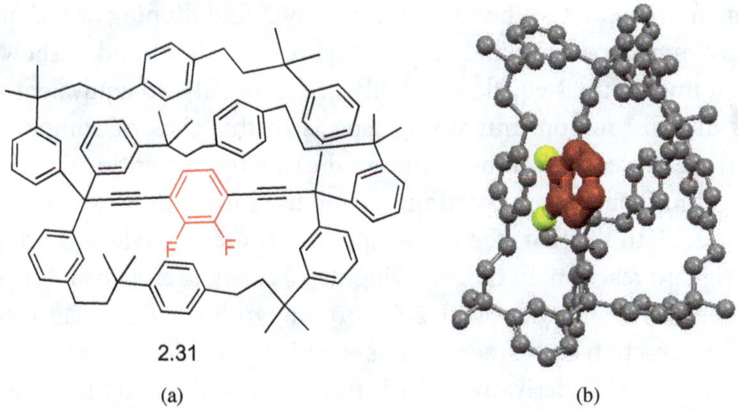

(a)

(b)

Figure 2.24. (a) An example of a rotor fitted to a cage like structure. (b) Ball and stick model to show the skeleton and rotor part is shown by space-filled model (hydrogen atoms are omitted for clarity of presentation).

2.3.3.6 *Molecular gyroscopes*

A rotor confined in a cage has similarity to a gyroscope; based on the model a general design principle is illustrated in Figure 2.25. For which three flexible units are to be connected to two common pivots where the rotor unit is connected (Figure 2.25). Several characteristic features are-needed in a molecule to behave as a rotor, three major characteristics

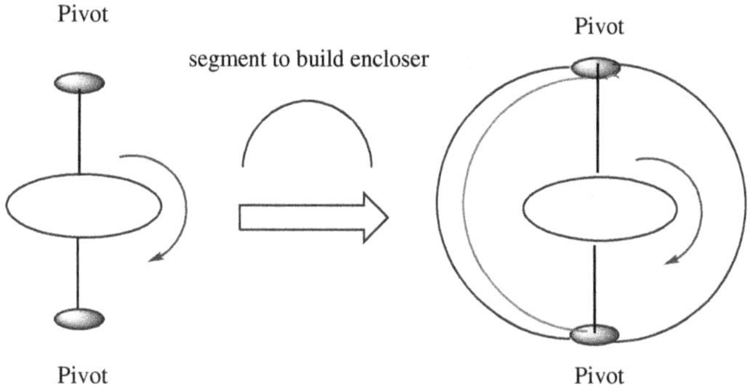

Figure 2.25. A design principle of a rotor in a closed environment.

relates to (a) Energy barrier for the rotation, (b) Size of the enclosure and (c) Interaction of rotor with the enclosure. As design part, enclosure has versatility by providing different weightage of flexibility, thus there is wide scopes to make cages in different manner.

Rotor part can be easily designed from a complex having trigonal bipyramid geometry having bulky ligands at axial positions such as in $Fe(CO)_3(PPh_3)_2$ (Figure 2.26). In this case the triphenylphosphine is a bulky ligand, hence acts as pivots for rotation of $Fe(CO)_3$ across Fe-P bonds. To modify this kind of rotor to a gyroscope, the rotor part has to be put in an enclosure. Coordinating atoms acting as pivots also decides the metal to pivot distance. This is one factor needed to successful design of such rotors. On the other hand, a rotor unit having a marker is easy for studies hence the effect of rotation can be easily seen in such gyroscope by using mixed ligand rotor. For example $Fe(CO)_2NO$ is used as rotating component in example **2.34**. Two other examples of gyroscopes prepared from arsenic based ligands with identical principle are given in Figure 2.27. These examples have $Fe(CO)_3$ unit enclosed in a cage-like structures is synthesized by metathesis reactions of olefin.[36] Hydrocarbon chains bearing olefinic end are connected to arsenic sites which are coordinated to iron. Unsaturated hydrocarbon cage is reduced by catalytic hydrogenation to form the molecular gyroscope **2.35c**.

Certain porous materials having low density in solid state have free volume, the free spaces provides avenues for motion of a component

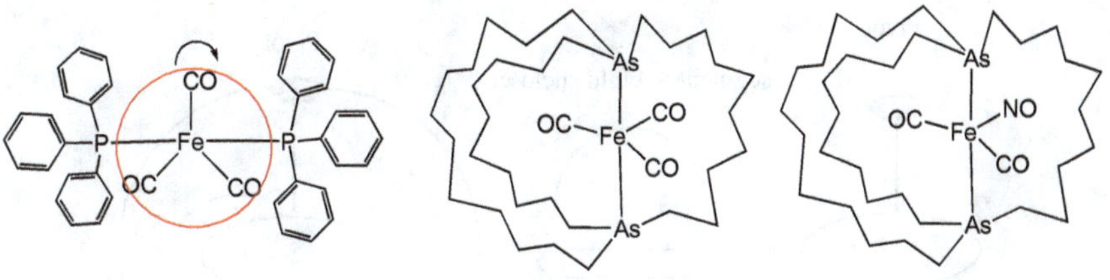

| 2.32 | 2.33 | 2.34 |
| (a) | (b) | (c) |

Figure 2.26. (a) Rotation of $Fe(CO)_3$ across Fe-P bonds, (b) and (c) are two complexes having property of a gyroscope.

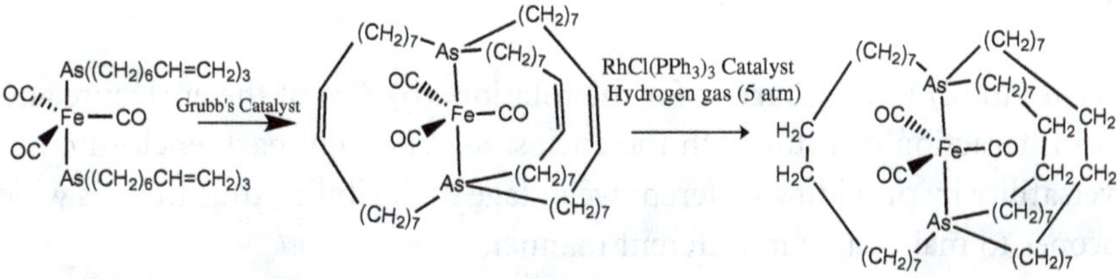

Figure 2.27. Synthesis of an arsenic ligand based gyroscopic molecule.

present within such spaces to show properties like molecular rotors. In such situation the life-time of many rotors falls in the range of nano-second. Such systems perform as rotors even at low temperature. Rotational barrier for such rotors can be overcome by diffusing gas molecules. Since the gas molecules are present as guests in voids of such materials, intake and let out of gas molecule can serve as fuel to run the rotors within the porous structure. In a solid state compounds are packed tightly but for having free rotation of a component some free volumes are essential. As a result of rotation of a component within a porous assembly or molecule an effect called "breath" occurs, this effect is nothing but the sorption and desorption of guest molecules.

There are also examples of rotors where motion created by electric field of dielectric materials causes rotational motion of the rotating part of a molecule. Inherent magnetic properties associated with them influences their mode and efficiency of performances. For example due to reorientation of dipoles by electric polarization a substance may have ferroelectric or anti-ferroelectric ground state. Above glass transition temperature a cross-linked polymer shows shape-persistent structure. A crystalline polymer showing such an effect executes directional movement in presence of stimuli.

2.3.3.7 *Double decker*

Based on a concept of having a rotatable part fixed on a stator, ruthenium based complex shown in Figure 2.28 has right kind of architecture and rotational features. This example is a complex **2.36** derived from ruthenium(II) ion with cyclopendienyl and hydrotris(indazolyl)borate ligands. Rigid platform in this compound is penta substituted cyclopentadienyl ligand. Substituents on cyclopentadienyl ligand are rigid and they

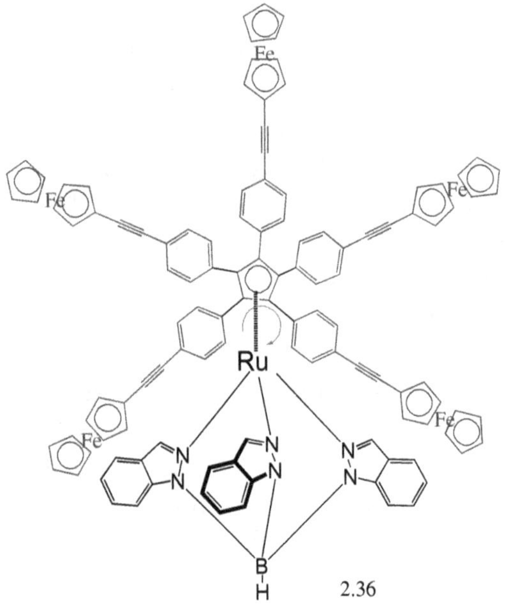

2.36

Figure 2.28. Ruthenium complex as a prototype of a molecular motor.

are like five arms attached to a pivot. The upper portion of this molecule has qualifications to rotate with respect to the lower portion constituted by ruthenium(II) hydrotris(indazolyl)borate.[37] Another, interesting aspect of this system is that, it has electro-active sites at terminal ends of the arms of the cyclyopentadienyl iron complexes as well as at the centrally located site that is ruthenium ion. Furthermore, cyclopentadiene iron complexes are well known for their one electron redox couple between +2 and +3 oxidation states. This part acts as reservoir for fuel that is received from an external electrical or chemical source. This molecule has also added advantage of having upper rotatable unit and a lower functionalisable base to fix to a surface. Such a modification or fixing to surface allows it to adopt locked conformation. There is also C-C bond rotation of the arms which are connected by single and triple bonds.

Molecular motor should have control on the speed of rotation that occurs in sequential steps or continuously. It should have also control over the direction of rotation. This is one point particularly essential in molecular systems comprising of multiple units as each units as its own characteristic. It is also advantageous to have critical understanding on such effects as in an assembly of molecules undergoing rotational motion of each part can be controlled by specific stimulus. Bis-porphyrinate double-decker zirconium and cerium complexes have features to control rotary speed (Figure 2.29). In these complexes metal center namely cerium or zirconium function as a ball-bearing as the metal ions are sandwiched between two porphyrin ligands (**2.38**). Porphyrin rings act like rotating disks over and below such balls. Having bulky substituents attached on the two porphyrin, chiral metal complexes are generated (Figure 2.30). The rotamers have different orientations of substituents are also mirror image conformations. One of the porphyrin ligand of the optically active complex rotates in opposite direction relative to another. Each optically active isomer shows independent rotary motion; where direction of rotary motion in each case is guided by the chirality of the isomer under consideration. Racemization occurs in this class of complexes by the rotation of optical isomers. Oxidation state of the metal center controls the rotation of these complexes. For example, relative rate of rotation is 300 times higher when the central cerium ion is in +4 oxidation state than

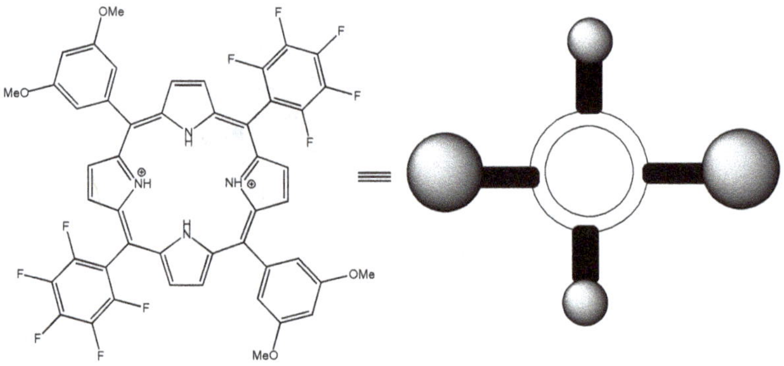

2.37

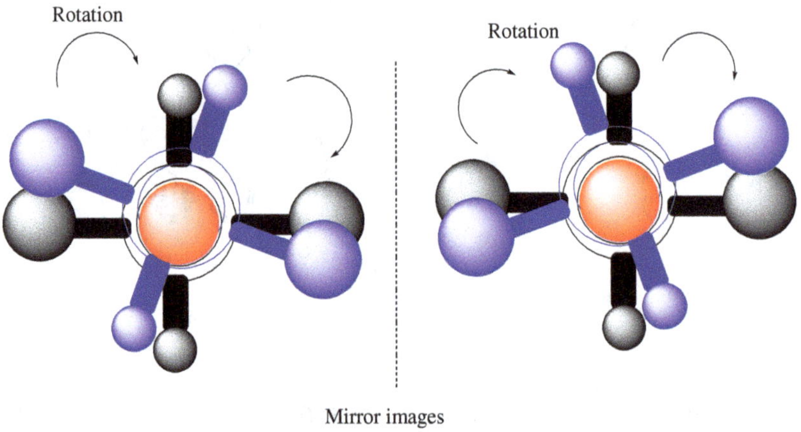

Mirror images

Figure 2.29. Rotational behavior of porphyrin ligand and chiral sandwich complex.[38] (Red ball is metal ion).

the complex having cerium ion in +3 oxidation state. On the other hand, similar zirconium complex shows a similar increase in rotation, but such enhancement in rotation occurs with zirconium complex at lower oxidation state. These changes in rate of rotation occur due to relative changes in ionic radius of metal ions. In higher oxidation state, ionic radius is short as well as the attractive force towards a negative charge reduces the rate of rotation of the ligand.

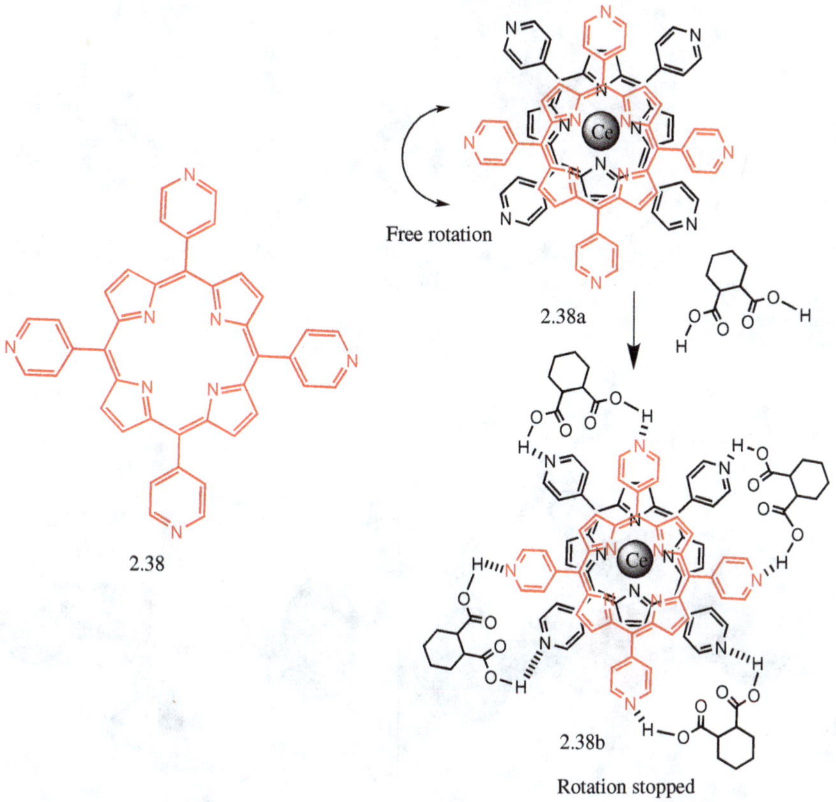

Figure 2.30. (a) A porphyrin derivative **2.38**, (b) Cerium complex of the derivative where free rotation over cerium ion is stopped by hydrogen bond formation with guest dicarboxylic acid molecules.[39]

2.3.3.8 *Internal rotations*

Diphenylsilica **2.39** has a structural feature as shown in Figure 2.31 has fixed siloxane as base to hold the diphenyl unit part by a distance of separation 4.4 Å. The diphenyl groups rotate about the Si-C bonds to show property of molecular rotor. Variable temperature ^{1}H and ^{13}C spin-lattice NMR relaxation times vary with temperature due to motion of phenyl rings. From such measurements spin-lattice relaxation times in ^{13}CNMR spectra at 75 MHz is maximum at 37°C. This indicates that this particular rotor respond in the 10^7 Hz. Similarly spin-lattice relaxation time recorded at 300 MHz and 30 MHz shows that rotational energy barrier is 6.4 Kcal/mol.

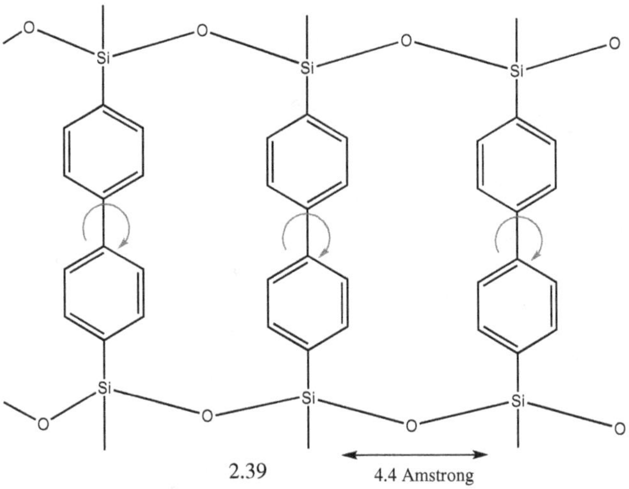

Figure 2.31. Structural backbone of diphenylsiloxane showing the diphenyl as rotating unit.

Organosilica based compounds are called mesoporous due to presence of cylindrical pores. Sizes of such pores are generally greater than 2 nm; these silicates are used to build integrated units with many rotors within them. To illustrate this let us consider *bis-trialkyloxy organosilane* which forms three siloxane bonds at each molecular end by replacing the three, alkoxy groups. Under suitable conditions this generates periodically aligned monomer units in the framework which have hexagonal arrangement with channel-like pores. In such architecture organic linkers are placed in ordered arrays favourable for rotation across each molecular axis. Orderly arrangements present in these structures of molecules make it possible to look for large numbers structural variations to study as molecular rotors. The length and nature of the rotating unit is varied by using different aromatic units such as phenyl, biphenyl etc in these structural frame-works (Figure 2.32). In this class of molecules siloxane parts act as spacers to keep the rotors apart from each other within the framework structure.

In metal organic frameworks also similar process of rotational motion of rotors take place. Rotation of phenyl groups are also observed in cubical structures constructed in metalloorganic frameworks. For example, metallo-organic frame work formed by zinc-1,4-dicraboxylate which is known as MOF-5 is such an example.

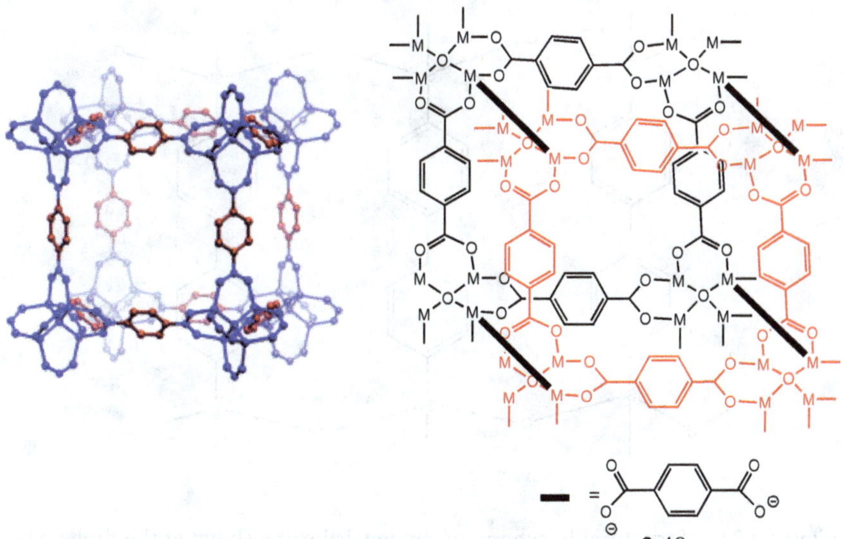

Figure 2.32. A portion of a cube-like structure of MOF-5 where phenyl groups show rotor action.[40]

Porphyrin rings connected by flexible tethers to make enclosure shown in Figure 2.33 may adopt face-to-face porphyrins or may be placed in oblique positions. Compound **2.41** shows fluxional behavior due to sliding of one porphyrin over another porphyrin at two different positions relative to each other. Adjustment of the position is guided by protonation and metal complexation.

Chiral carcerand **2.42** shows optical isomers in the form of enantiomer or may adopt optically inactive meso form (Figure 2.34). The shape of such molecules can be modulated by solvent as depending on interacting solvent surface area can vary leading to mechanical motion in different direction with change of dielectric.

2.3.3.9 *Competitive ion binding*

Competitive binding of cations with crown ether like structures helps in expansion and contraction in a twisted manner. For example, ligand **2.42a** forms sodium salt of boron complex **2.42b**. In this complex

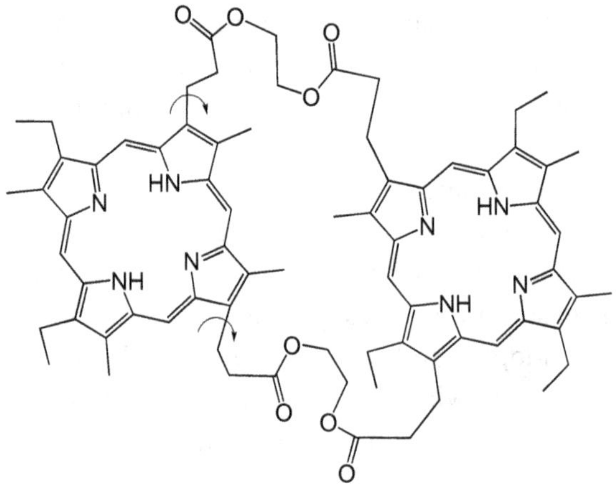

2.41

Figure 2.33. Stimuli responsive flexible porphyrin dimer to adjust positions of porphyrin rings.[41]

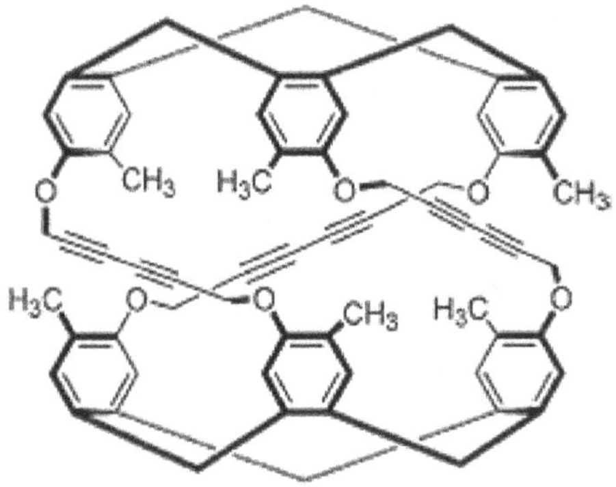

2.42

Figure 2.34. A carcenand for possibility to modulate surface area.[42]

sodium ion gets encapsulated in the anionic part. This is due to a cyclic crown like environments of oxido groups in the complex. Oxido groups interact with sodium ion to form a compact structure. In this compact structure the distance between the boron atoms is 6 Å. Encapsulated sodium ion from this compound is brought outside by a cryptand to

form **2.42c**. This happens due to high affinity of the cryptand for sodium
ion. In this process, sodium ion gets encapsulated into the cage of the
cryptand and boron complex anion is devoid of sodium ion that was initially concealed by it. Due to absence of a cation inside the structure **2.42c**,
thus complex anion expands through displacement of positions of boron
atoms. Distances between the boron atoms get adjusted showing change of
distance from 6 Å longitudinal distance to such distance of separation to 13
Å. This change in structure process also involves puckering of molecules as
the boron atoms adopts tetrahedral environment (Figure 2.35). This con-
strained by interactions with sodium ions cause contraction in size hence
differs in size, hence the one which environment without sodium ion.

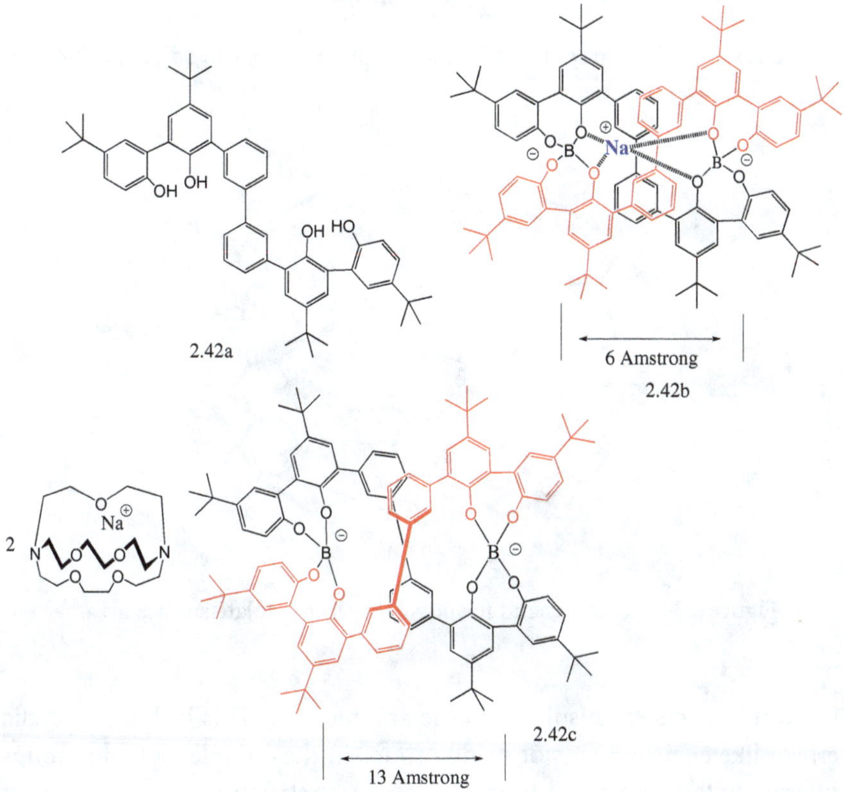

Figure 2.35. Contraction and expansion of boron complex through sodium ion
encapsulation[43]

2.3.3.10 *Modulation by hydrogen bonds*

2,6-Pyridinecarboxaldiamides are semi-rigid due to presence of amide functional groups (Figure 2.36). They adopt various conformers by C-C rotation, whereas on protonation of the pyridine part of such compounds the neighboring carbonyl groups hydrogen bonds to N-H bond and it makes rigid structure (**2.43a**). Whereas, opposite situation arises in the case of 2,6-diaminopyridine (**2.44a**), it adopts rigid structure due to hydrogen bond, whereas the protonation of the nitrogen atom of pyridine it becomes rigid. These two types of compounds differ in acidity and the relative ability to get protonated differs. Advantages of such processes are taken to construct equilibration between linear and spiral structures.

Polyamide compound **2.46a** shown in Figure 2.37 has **2.43a** and **2.44a** as constituent components in alternating positions. Due to rotation of the C-C bond next carbonyl groups, compound in neutral condition is spiral and due to steric reasons it adopts thermodynamically stable structure. Protonation of nitrogen on the ring of the pyridine flanked by carbonyl by a mild acid, such molecules flatten. Flattening is due to formation of hydrogen bonds involving each nitrogen atoms of pyridine rings in such compounds. Upon treatment with strong acid all pyridine nitrogen atoms of the molecule **2.45** are protonated. This provides scope to get relieved of original intramolecular hydrogen bonds of the pyridine rings with two

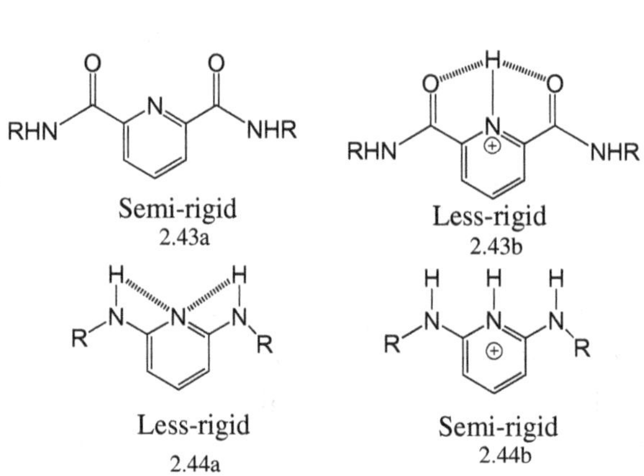

Figure 2.36. Intramolecular hydrogen bonds in 2,6-pyridinecarboxaldiamides and 2,6-diaminopyridinederivatives.

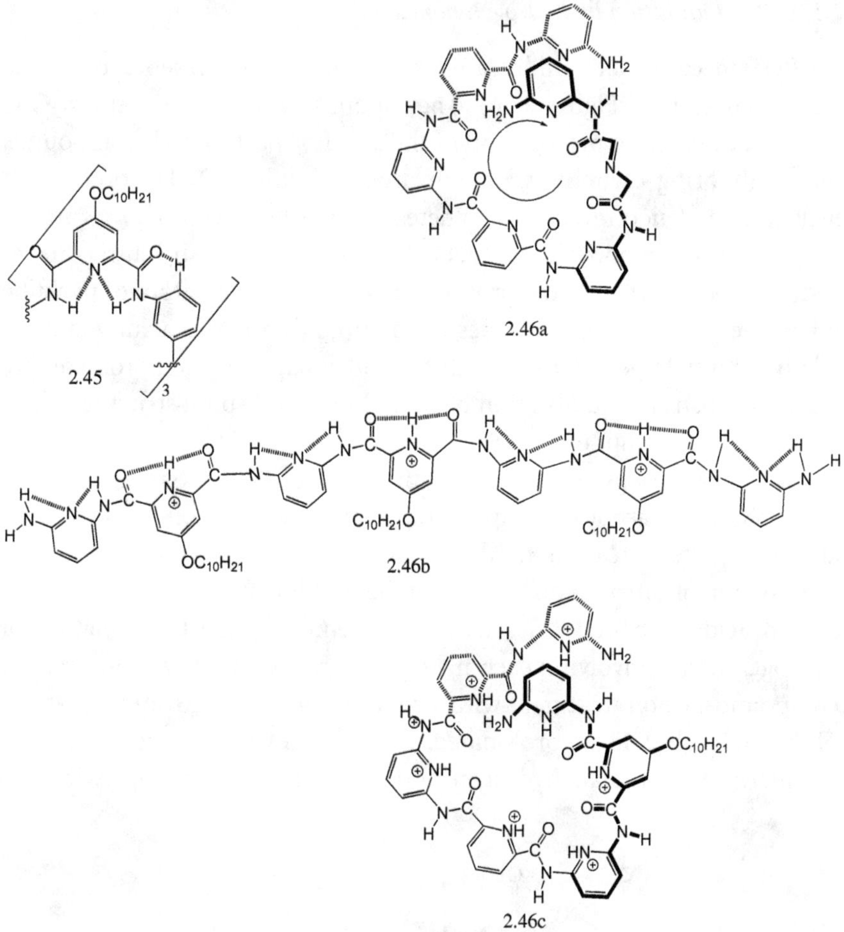

Figure 2.37. Modulation of linear and spiral structure though hydrogen-bonds.[44]

amide groups. Hence, rigidity on pyridinium ring flanked by carbonyl is retained but the carbon-nitrogen bond rotation of the pyridinium ion flanked by amino groups is set on. This process generates a puckered structure through rotation of C-N bonds to attain a spiral structure **2.46b** which is different from helical structure **2.46a** that is observed in neutral state. Thus, depending on concentration of hydrogen ions these three states equilibrate in solution. Such changes induced by change of hydrogen bonds result in expansion and contraction of the molecule providing mechanical motion.

Equilibration between stretched and spiral states can also be brought about by metal coordination of a spiral poly-terpyridine ligand. Ligand **2.47a** has 2, 2′-bipyridine units in alternate positions. Since anti-geometry provides the stable conformation of 2, 2′-bipyridine; molecule has a spiral structure and it adopts geometry **2.47a** as shown in Figure 2.38. The structure of the compound is linear (**2.47c**) when it coordinates to six lead ions. In the complex with lead ionsnitrogen atoms lie on one side of the structure to provide a linear structure. Thus, compound **2.47a** on complex formation with lead ions straightens and on decomplexation goes back to spiral structure.

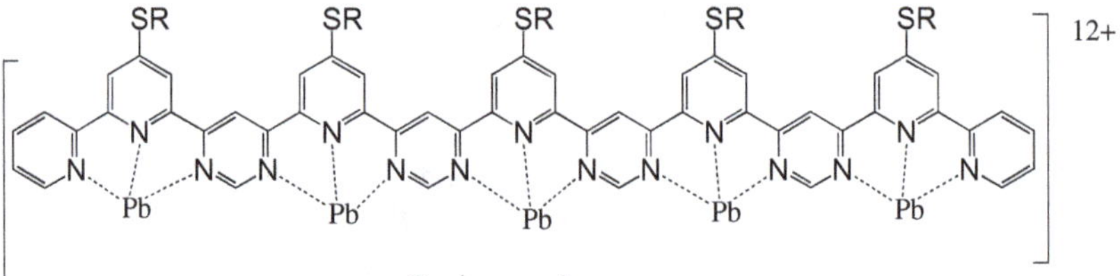

R = isopropyl

2.47c

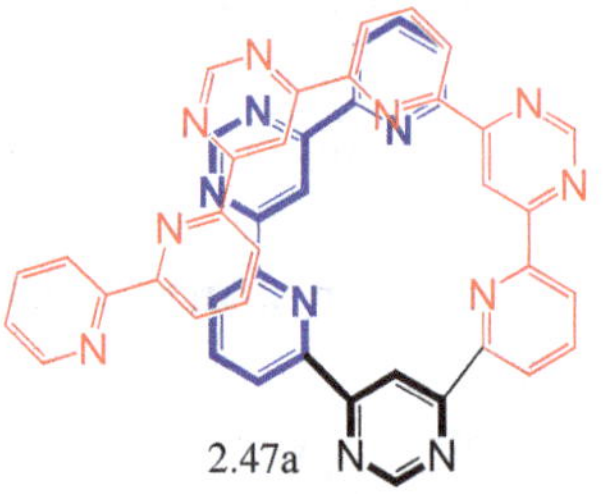

2.47a

Stable folded structure; SR group omitted for clarity

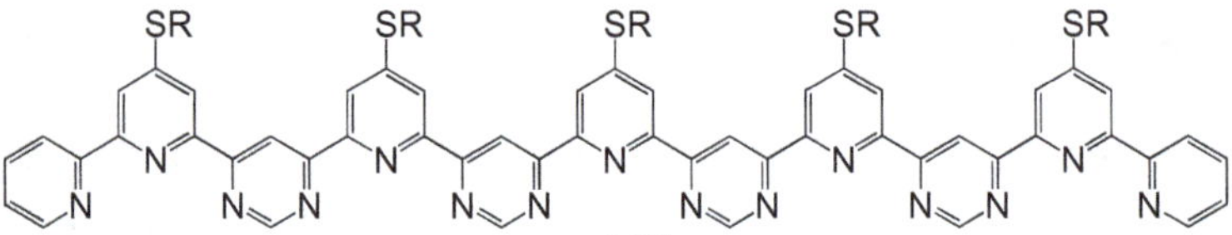

2.47b

Linear structure thermodynamically unfavorable

Figure 2.38. Coordination of lead ion results in stretched structure which was originally helical and equilibration results in expansion and contraction.[45]

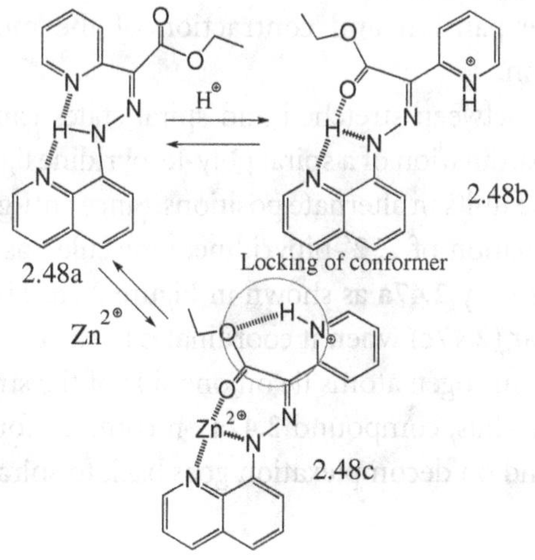

Figure 2.39. Locking of equilibrium by protonation.[46]

A compound having multiple binding sites for acid binding, metal complexation, the geometry of different kinds may be equilibrated by using these properties (Figure 2.39). For example, 8-azaquinoline deriva- tive **2.48a** has a pyridine ring which participates in hydrogen bonds to form a structure where carbonyl group is outward from aromatic rings. When **2.48a** is treated with acid, pyridine gets protonated and hydrogen bond acceptor property of pyridine is lost. Accordingly, N•••H bond breaks down and a new C=O•••H bond is formed. Under such a situation, pyri- dinium cation adopts a new geometry observed in structure **2.48b**. In this structure nitrogen atoms project inward with respect to rest part of the molecule. Upon complexation to zinc ion, molecule brings the pyridinium ion towards inward direction. Thus it gets stabilised in the form **2.48c** by forming intra-molecular hydrogen bonds between oxygen atom of ester and N-H bond of pyridinium cation.

Variation in pH and complex formation with zinc ion is utilized as dual roles to bring about structural changes and control proton transfer in pair compounds **2.49a** and **2.50a**. These are two compounds that possess intra-molecular hydrogen bonds (Figure 2.40). The compound **2.49a** on reaction with zinc ion form a zinc complex **2.49b**, which triggers the imi- dazolium cation to occupy an orientation at the opposite side of quinolone

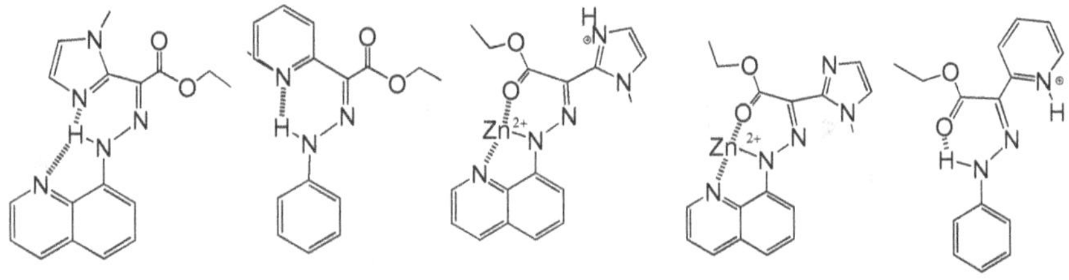

| 2.49a | 2.50a | 2.49b | 2.49c | 2.50b |

Figure 2.40.   Locking of conformers by pH and zinc complexation and proton transfer.

ring changing the conformation from the free ligand. This zinc complex **2.49b** transfer proton to compound **2.50a** and thereby itself changes to **2.49c**. This allows the pyridine part of **2.50a** gets unlocked and carbonyl group replaces position of the pyridine to form new intramolecular hydrogen bonds involving carbonyl oxygen atom with N-H of hydrazine part of the molecule. Thus, there is change of conformation of **2.50a** to **2.50b** by proton transfer. This change is caused by formation of zinc complex. Conversion of **2.49a** to **2.49c** during the process involves conformational change through complexation, protonation and deprotonation process. Entire process causes mechanical changes in these two molecules under different conditions and stimuli.

2.3.3.11 *Charge-transfer*

Non-covalent inter-molecular interactions and charge-transfer interactions are used to tune molecular arrangements. Polyaromatic hydrocarbons with D_{6h}-symmetry have special interest due to low rotational barrier and they serve as models for fragmented graphene. Such polyaromatic hydrocarbons form charge-transfer complexes with electron deficient compounds such as tetrafluoro 7,7,8,8 tetracyanoquinodimethane (F4TCNQ). Charge-transfer complex of coronene with F4TCNQ namely (coronene)(F4TCNQ) shown in Figure 2.41 exhibits slow in-plane molecular rotation. This rotation is due to formation of a new packing pattern with A-D-D-A-D-D-A type of stacking arrangements from the original packing pattern having D-A-D-A arrangement. Multiple donor-donor-acceptor type of packing

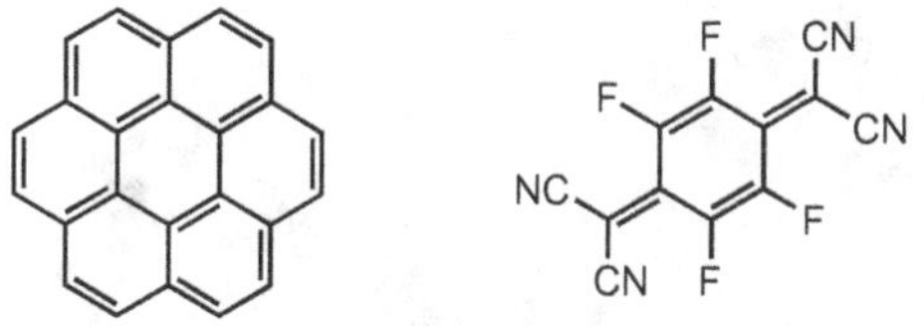

Coronene   tetrafluoro 7,7,8,8 tetracyanoquinodimethane

D = Donor   A = Acceptor

Figure 2.41. Donor acceptor pairs that form different charge-transfer complexes.

Gold surface

(a)   (b)

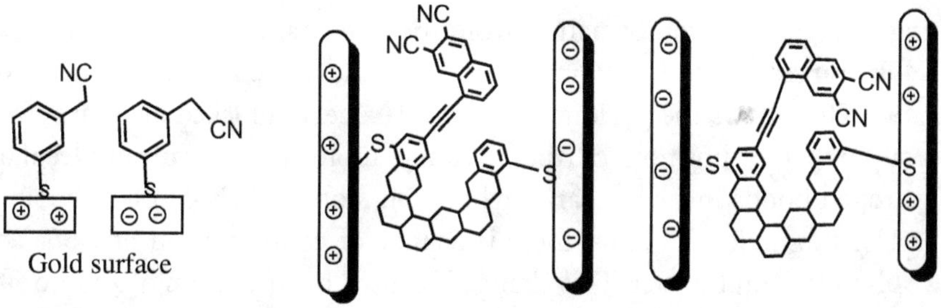

Figure 2.42. Examples of conformational molecular rectifiers: (a) due to different orientation of the cyano-group and (b) Different conformation due to change in orientation of naphthyl group.[48]

arrangements formed by transformation is less stable but due to small energy differences the packing preferences of particular structure such interchanges is possible. In solid-state these charge-transfer molecules undergo in-plane rotation of coronene molecules in giga-hertz frequency. Depending on molecules used in charge-transfer complexes dynamic properties of such systems vary.[47]

Conformational molecular rectifiers (CMR) of molecules attached to gold surfaces through their thiol groups are shown in Figure 2.42a and Figure 2.42b. These molecules are suitable for changing conformation once placed in a tunneling junction. Characteristic current-voltage plots of these molecules show significant rectification. In each examples shown in Figure 2.42, molecules are comprise of a dipolar rotating part whose orientations are dependent on electric field.

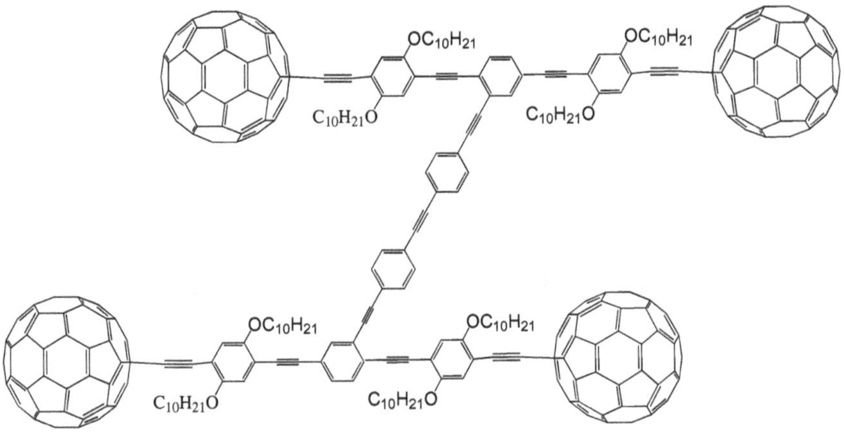

Figure 2.43. A nano-car.

2.3.3.12 *Nano-car*

A four-wheeled "nanocar" with fullerene wheels[49–50] is shown in Figure 2.43. This molecule moves on an Au(111) surface when thermal energy is supplied. Translation motion is directional and it moves perpendicular to the axles of the molecule. This movement suggests that there is rolling motion of the fullerene wheels and the movement is not a sliding motion of the molecule that would have caused movement in all direction of the surface. Such movement also can be caused by using a scanning tunnelling microscope tip. Scanning tunnelling microscope is an instrument that is used for imaging surfaces at atomic level with lateral resolution of 0.1 nm and depth resolution 0.01 nm.

2.3.3.13 *Molecules walking*

Competitive reactions and reversible reactions are utilized to move a molecule along a direction. This is done by causing shift of positions of a molecule or molecular entity towards a particular direction. Figure 2.44 is a description of such a walker system. The molecule described here has two chemically different legs and these legs are labile under orthogonal conditions. The molecule at terminal has a hydrolysable hydrazone group and a disulphide link making a cyclic structure. Cyclic structure is built over aromatic rings. Aromatic rings serve as support to moving part of the

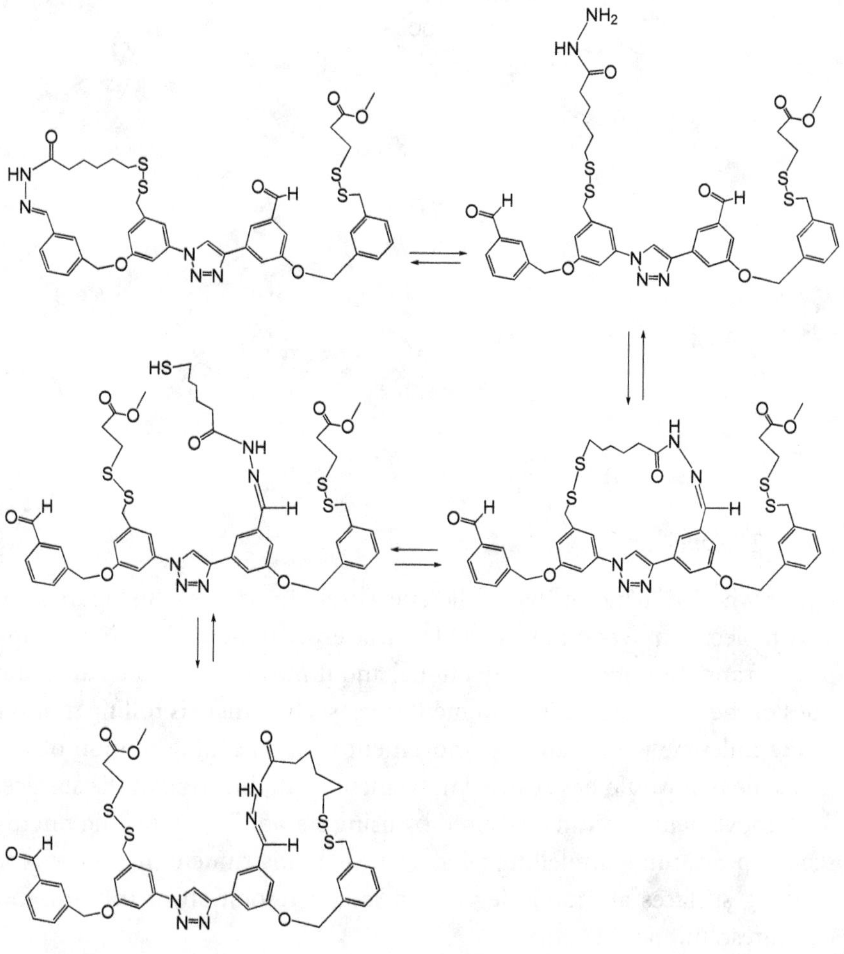

Figure 2.44. A small molecule as walker, walking along a molecular track.[51]

molecule. There are two other identifiable positions of aromatic rings on the chain; one ring has an aldehyde group and next one has disulphide linked ester part. Hydrolysis of hydrazone causes shift of a part of the molecule from the cyclic unit at one terminal to the next aromatic ring to form free hydrazine unit. Hydrazide thus generated reacts with aldehyde group located at next position to form again a new cyclic structure. This cyclic structure again opens up and another cyclic structure is formed at the cost of opening of earlier ring. Overall process is like somersaults motion of a molecule from one end to another end.

References

1. Peplow M (2015) March of the machines. *Nature* **525**: 18–21.
2. Wang G, Ma L, Xiang J, Wang, Y, Chen X, Che Y, Jiang H (2015) 2,6-Pyridodicarboxamide-bridged triptycene molecular transmission devices: converting rotation to rocking vibration. *J. Org. Chem.* **80**: 11302–11312.
3. Wang G, Xiao H, He J, Xiang J,Wang Y, Chen X, Che Y, Jiang H (2016) Molecular turnstiles regulated by metal ions. *J. Org. Chem.* **81**: 3364–3371.
4. Karlen SD, Garcia-Garibay MA (2005) Amphidynamic crystals: structural blueprints for molecular machines. *Top. Curr. Chem.* **262**: 179–227.
5. Vives G, Jacquot de Rouville H-P, Carella A, Launa J-P, Rapenne G (2009) Prototypes of molecular motors based on star-shaped organometallic ruthenium complexes. *Chem. Soc. Rev.* **38**: 1551–1561.
6. Sanada K, Ube H, Shionoya M (2016) Rotational control of a dirhodium-centered supramolecular four-gear system by ligand exchange. *J. Am. Chem. Soc.* **138**: 2945–2948.
7. Duyker SG, Peterson VK, Kearley GJ, Studer AJ, Kepert CJ (2016) Extreme compressibility in $LnFe(CN)_6$ coordination framework materials via molecular gears and torsion springs. *Nature Chem.* **8**: 270–275.
8. Tashiro K, Konishi K, Aida T (2000) Metal bisporphyrinate double-decker complexes as redox-responsive rotating modules. Studies on ligand rotation activities of the reduced and oxidized forms using chirality as a probe. *J. Am. Chem. Soc.* **122**: 7921–7926.
9. Nishiyama Y, Inagaki Y, Yamaguchi K, Setaka W (2015) 1,4-Naphthalenediyl-bridged molecular gyrotops: rotation of the rotor and fluorescence in solution. *J. Org. Chem.* **80**: 9959–9966.
10. Vogelsberg CS, Garcia-Garibay MA (2012) Crystalline molecular machines: function, phase order, dimensionality, and composition. *Chem. Soc. Rev.* **41**: 1892–1910.
11. Dominguez Z, Dang H, Strouse MJ, Garcia-Garibay MA, (2002) Molecular "Compasses" and "Gyroscopes" expedient synthesis and solid state dynamics of an open rotor with a bis(triarylmethyl) frame. *J. Am. Chem. Soc.* **124**: 2398–2399.
12. Setaka W, Yamaguchi K (2013) Order-disorder transition of dipolar rotor in a crystalline molecular gyrotop and its optical change. *J. Am. Chem. Soc.* **135**: 14560–14563.

13. Rodriguez-Molina B, Ochoa ME, Farfan N, Santillan R, Garcia-Garibay MA (2009) Synthesis, characterization, and rotational dynamics of crystalline molecular compasses with N-heterocyclic rotators. *J. Org. Chem.* **74**: 8554–8565.

14. Wong KL, Pawin G, Lin X-Y, Jiao T, Solanki U, Fawcett RHJ, Bartels L, Stolbov S, Rahman TS (2007) A molecule carrier, *Science* **315**: 1391–1392.

15. Hrovat DA, Borden WT, Eaton PE, Kahr B (2001) A computational study of the interactions among the nitro groups in octanitrocubane. *J. Am. Chem. Soc.* **123**: 1289–1293.

16. Tashiro K, Konishi K, Aida T (2000) Metal bis-porphyrinate double-decker complexes as redox-responsive rotating modules. Studies on ligand rotation activities of the reduced and oxidized forms using chirality as a probe. *J. Am. Chem. Soc.* **122**: 7921–7926.

17. Engeser M, Fabbrizzi L, Licchelli M, Sacchi D (1999) A fluorescent molecular thermometer based on the nickel(II) high-spin/low-spin interconversion. *Chem. Commun.* 1191–1193.

18. Conte M, Hippler M (2016) Dynamic NMR and quantum-chemical study of the stereochemistry and stability of the chiral $MoO_2(acac)_2$ complex in solution. *J. Phys. Chem. A* **120**: 6677–6687.

19. Wang XB, Dai B, Woo HK, Wang LS (2005) Intramolecular rotation through proton transfer: $[Fe(\eta5-C_5H_4CO_2^-)_2]$ versus $[(\eta^5-C_5H_4CO_2-)Fe(\eta^5-C_5H_4CO_2H)]$. *Angew. Chem., Int. Ed.* **44**: 6022–6024.

20. Dial BE, Pellechia PJ, Smith MD, Shimizu KD (2012) Proton Grease: An Acid Accelerated Molecular Rotor. *J. Am. Chem. Soc.* **134**: 3675–3678.

21. Hrovat DA, Borden WT, Eaton PE, Kahr B (2001) A Computational study of the interactions among the nitro groups in octanitrocubane. *J. Am. Chem. Soc.* **123**: 1289–1293.

22. Karmakar A, Sarma RJ, Baruah JB (2007) Polymorphism in an aqua-bridged, dinuclear 2-nitrobenzoate complex of cobalt(II). *Eur. J. Inorg. Chem.* 643–647.

23. Kottas GS, Clarke LI, Horinek D, Michl J (2005) Artificial Molecular Rotors. *Chem. Rev.* **105**: 1281–1376.

24. Horinek D, Michl J (2005) Surface-mounted altitudinal Molecular rotors in alternating electric field: single-molecule parametric oscillator molecular dynamics. *Proc. Natl. Acad. Sci. U. S. A.* **102**: 14175–14180.

25. Jian H, Tour JM (2003) En Route to Surface-Bound Electric Field-Driven Molecular Motors. *J. Org. Chem.* **68**: 5091–5103.

26. Asfari Z, Naumann C, Kaufmann G, Vicens J (1998) Synthesis of a molecular mill designed from a calix[4]-*bis*-crown. *Tetrahedron Lett.* **39**: 9007–9010.

27. Iwamura H, Mislow K (1988) Stereochemical consequences of dynamic gearing. *Acc. Chem. Res.* **21**: 175–182.

28. Kelly TR, Bowyer MC, Bhaskar KV, Bebbington D, Garcia A, Lang F, Kim MH, Jette MP (1994) A Molecular Brake. *J. Am. Chem. Soc.* **116**: 3657–3658.

29. Rebek J, Trend JE, Wattley RV, Chakravorti S, (1979) Allosteric effects in organic chemistry: site-specific binding. *J. Am. Chem. Soc.* **101**: 4333–4337.

30. Setaka W, Nirengi T, Kabuto C, Kira M (2008) Introduction of clutch function into a molecular gear system by silane-silicate interconversion. *J. Am. Chem. Soc.* **130**: 15762–15763.

31. Kelly TR, De Silva H, Silva RA (1999) Unidirectional Rotary motion in a molecular system. *Nature* **401**: 150–152.

32. Petitjean A, Khoury RG, Kyritsakas N, Lehn J-M (2004) Dynamic devices: shape switching and substrate binding in ion-controlled nanomechanical molecular tweezers. *J. Am. Chem. Soc.* **126**: 6637–6647.

33. Ng PL, Lambert JN (1999) Synthesis of Macrocyclic Inclusion Complexes Using Olefin Metathesis. *Synlett* 1749–1750.

34. Dominguez Z, Khuong T-AV, Dang H, Sanrame CN, Nunez JE, Garcia-Garibay MA (2003) Molecular Compasses and Gyroscopes with Polar Rotors: Synthesis and Characterization of Crystalline Forms. *J. Am. Chem. Soc.* **125**: 8827–8837.

35. Godinez CE, Zepeda G, Garcia-Garibay MA (2002) Molecular compasses and gyroscopes. II. Synthesis and characterization of molecular rotors with axially substituted bis[2-(9-triptycyl)ethynyl]arenes. *J. Am. Chem. Soc.* **124**: 4701–4707.

36. Lang GM, Bhuvanesh N, Reibenspies JH, Gladysz JA (2016) Syntheses, reactivity, structures, and dynamic properties of gyroscope-like iron carbonyl complexes based on dibridgehead diarsine cages. *Organometallics* **35**: 2873–2889.

37. Cardella A, Jaud J, Rapenne G, Launay J-P (2003) Technomimetic molecules: synthesis of ruthenium(II) 1,2,3,4,5-penta(p-bromophenyl)cyclopentadienyl hydrotris- (indazolyl)borate, an organometallic molecular turnstile *Chem. Commun.* 2434–2435.

38. Tashiro K, Konishi K, Aida T (1997) Enantiomeric resolution of chiral metallobis(porphyrin)s: Studies on rotatbility of electronically coupled porphyrin ligands. *Angew. Chem. Int. Ed.* **36**: 856–858.

39. Shinkai S, Ikeda M, Sugasaki A, Takeuchi M (2001) Positive Allosteric Systems Designed on Dynamic Supramolecular Scaffolds: Toward Switching and Amplification of Guest Affinity and Selectivity. *Acc. Chem. Res.* **34**: 494–503.

40. Zhu K, O'Keefe CA, Vukotic NV, Schurko RW, Loeb SJ (2015) A molecular shuttle that operates inside a metal–organic framework. *Nature Chem.* **7**: 514–519.

41. Leblond J, Petitjean A (2011) Molecular Tweezers: Concepts and Applications. *ChemPhysChem* **12**: 1043–1051.
42. Cram DJ, Tanner ME, Kiepert SJ, Knobler CB (1991) Two Chiral [1.1.1] Orthocyclo-phane Units Bridged by Three Biacetylene Units as a Host Which Binds Medium-Sized Organic Guests. *J. Am. Chem. Soc.* **113**: 8909–8916.
43. Miwa K, Furusho Y, Yashima E (2010) Ion-triggered spring-like motion of a double helicate accompanied by anisotropic twisting. *Nature Chem.* **2**: 444–449.
44. Tian F, Jiao DZ, Biedermann F, Scherman OA (2012) Orthogonal switching of a single supramolecular complex. *Nature Commun.* **3**: 1207.
45. Barboiu M, Lehn J-M (2002) Dynamic chemical devices: modulation of contraction/extension molecular motion by coupled ion binding/pH change-induced structural switching. *Proc. Natl. Acad. Sci. U. S. A.* **99**: 5201–5206.
46. Langde SM, Aprahamian I (2009) A pH activated configurational rotary switch: Controlling the E/Z isomerization in hydrazones. *J. Am. Chem. Soc.* **131**: 18269–18271.
47. Yoshida Y, Kumagai Y, Mizuno M, Isomura K, Nakamura Y, Kishida H,G. Saito G (2015) Improved dynamic properties of charge-transfer-type supramolecular rotor composed of coronene and F_4TCNQ. *Cryst. Growth Des.* **15**: 5513–5518.
48. Troisi A, Ratner MA (2004) Conformational molecular rectifiers. *Nano Lett.* **4**: 591–595.
49. Morin J-F, Shirai Y, Tour JM (2006) En route to a motorized nanocar. *Org. Letter.* **8**: 1713–1716.
50. Shirai Y, Osgood AJ, Zhao Y, Kelly KF, Tour JM (2005) Directional control in thermally driven single-molecule nanocars. *Nano Lett.* **5**: 2330–2334.
51. Yin P, Yan H, Daniell XG, Turberfield AJ, Reif JHA (2004) Unidirectional DNA walker that moves autonomously along a track. *Angew. Chem. Int. Ed.* **43**: 4906–4911.

3

Interlocked Systems as Molecular Machines

3.1 Principles for Formation of Interlocked Molecules

Activation of mechanical motions on a molecular system or self-assembly is generally guided by weak interactions. To understand such effects one may consider the ability of crown ether to recognize ammonium cation through hydrogen bonds.[1–2] This effect can be utilized to stabilize different optical isomers. Optically active crown ether **3.1a** recognizes optically active isomers of the PhCHMe-NH$_3^+$PF$_6^-$ (**3.1c**) (Figure 3.1). Energy difference between the two optical isomers of **3.1c** is 0.3 kcal/mol. But slight difference upon host-guest complex formation is observed. This is due to difference in environment while binding two isomers by host **3.1a**. One optical isomer of **3.1c** has methyl group which witness repulsion from neighbouring hydrogen atoms as illustrated by green arrow in Figure 3.1e. Such repulsion is not present in the host-guest complex of the other isomer. Hence the proportion of two forms in solution is 62 : 38 favouring form **3.1d** due to steric reason. On the other hand, host crown ether **3.1b** shows such ratio 97 : 3. This is due to difference in scaffolds of two crown ether hosts. In later case host is more compact hence repulsion is more prominent in one form. Electronic effects of the phenylene group containing crown ethers or cationic part in cyclophane host are utilized to encapsulate electron deficient or electron rich guests **3.2a-b** as per principle shown in Figure 3.2. Cryptate are class of compounds, where more ring/s is/are constructed over one ring keeping constant connecting points for the rings constructed over other. Thus, they have cage like enclosures

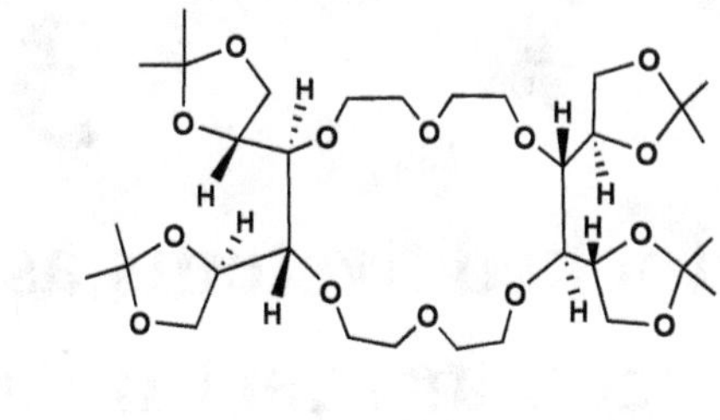

3.1a 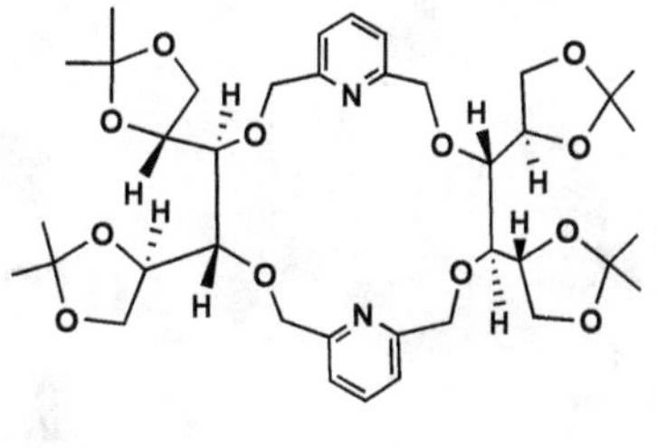   3.1b

[PhCH(Me)NH₃]PF₆

3.1c

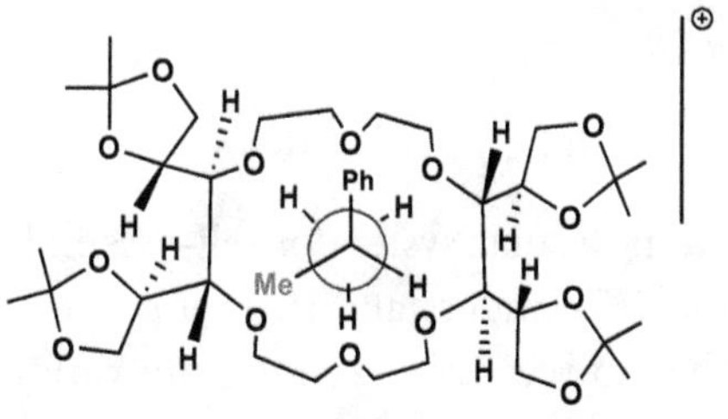

3.1d 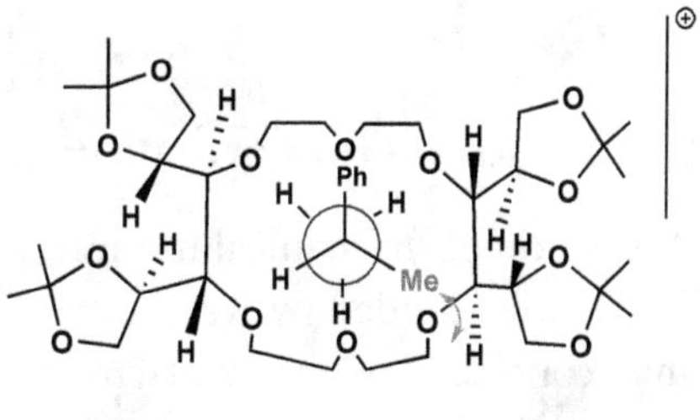   3.1e

Figure 3.1. **3.1a** and **3.1b** are two host crown ethers that bind with chiral [PhCH(CH₃)NH₃]PF₆ (**3.1c**). **3.1d** and **3.1e** are host-guest complexes of crown ether **3.1a** with the chiral salt.

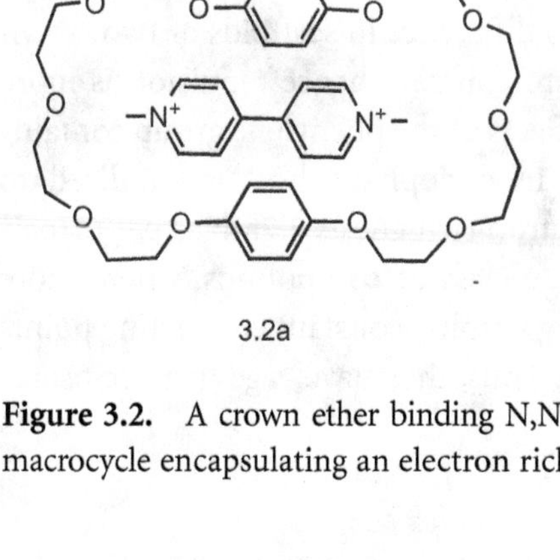

3.2a 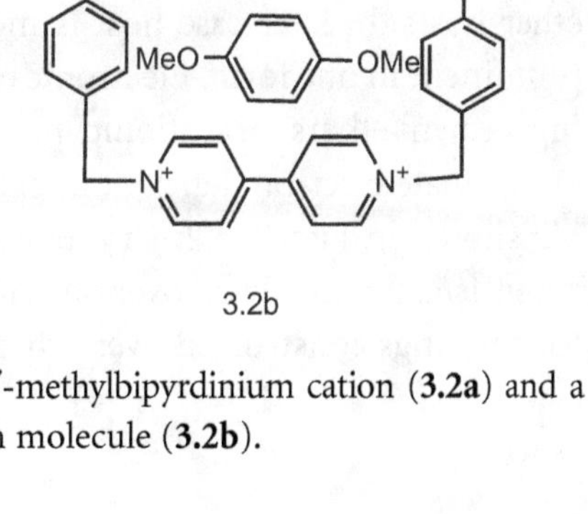   3.2b

Figure 3.2. A crown ether binding N,N'-methylbipyrdinium cation (**3.2a**) and a macrocycle encapsulating an electron rich molecule (**3.2b**).

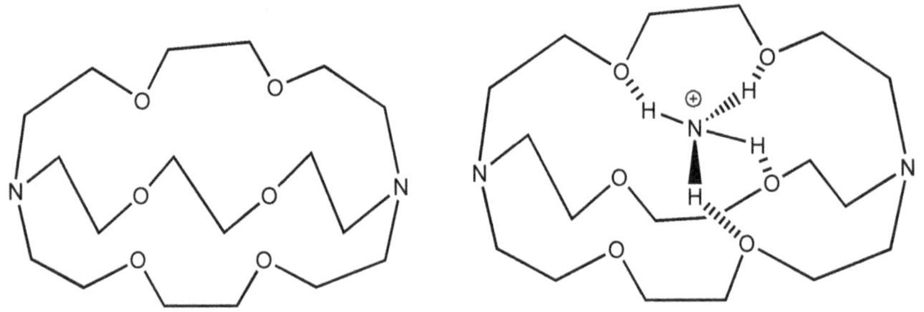

3.3a   3.3b

Figure 3.3.   A cryptand **3.3a**, and cryptand encapsulating an ammonium cation (**3.3b**).

to accommodate guest molecules or ions. Ammonium cation containing cryptate **3.3a** (Figure 3.3) has a rotational barrier for rotation about a C_3 axis i.e. to make a 120° rotation about an axis passing through nitrogen atom of the ammonium cation is 23 kcal mol^{-1}.

Such ammonium cryptate belongs to a class of compounds called as anisodynamic supramolecule. Compound has a suitable geometry for formation of hydrogen bonds by remaining within the cavity. Thus, an ammonium cation easily gets encapsulated within the cryptand. ^{1}HNMR of ammonium cryptate at room temperature shows fluxional behavior. In this particular case fluxional behavior is not seen below 228 K. At this temperature motions get frozen. The process takes place in solution due to symmetric structure formed by exchange of positions of the N-H···O hydrogen bonds of ammonium cation with the ethereal oxygen atoms. Such exchanges do not take place in crystalline compound. In solid state ammonium cation makes a rigid structure by getting encapsulated by the host. Energy of activation required for the exchange process in solution to cause symmetrization of the host-guest complex is 10.5 kcal mol^{-1}.

Forgoing examples explain encapsulation of guests in which guest molecules can be dislodged readily. Examples are host-guest systems that are purely guided by non-covalent interactions. Molecules can be also interlocked without bonds so that a stimulus can cause movement among them or displace them from one site to another by providing a defined path. Generally, inter-locked molecules are comprised of two or more mechanically-linked components between which no covalent bonds exist. Large numbers of molecular machines are designed from interlocked organic molecular compounds with distinguishable shapes. Most common families of interlocked molecular compounds are catenanes, rotaxanes and knots. These structures are suitable for mechanical movements by stimuli; hence many of them act as molecular machines. Shape and supramolecular properties guide their performance and control.

3.2 Catenanes

Catenanes consist of at least two interlocked macrocycles (Figure 3.4). When interlocked macrocyclic rings have two different recognition sites, then equilibration between such positions execute dynamic processes which is like movements in molecular shuttles. In such situations there is possibility for reversible equilibration between different sites that may be called as switching on and off. When interlocked rings are unsymmetric then recognition, building binding and stimuli response properties differs. Depending on differences in recognition properties of any of recognition sites, it is possible to induce conformational changes. Conformational changes among the rings can be viewed as the rotation of the non-symmetric ring. When symmetric interlocked rings are used, they have similar recognition sites. Under this circumstance, one of the rings may rotate inside the other among multiple numbers of rings. Such rotations are possible when non-covalent interactions, most commonly hydrogen bond stabilize particular conformations. In such a situation two states created in the inner ring, are indistinguishable and are called as degenerate states. When interlocked rings have different sites which respond independently

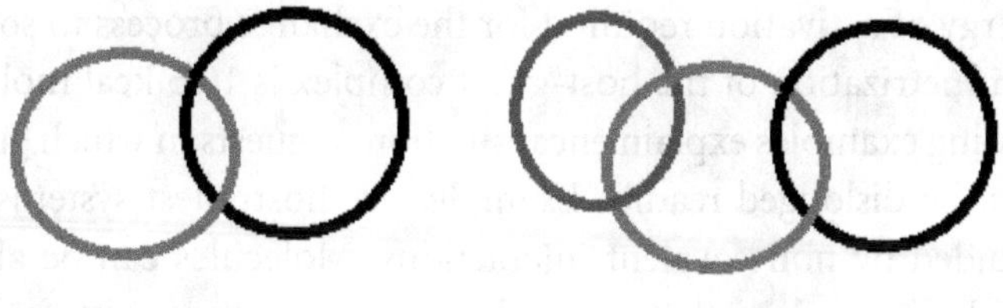

3.4(a)   3.4(b)

Figure 3.4.   A schematic representation of interlocked rings.

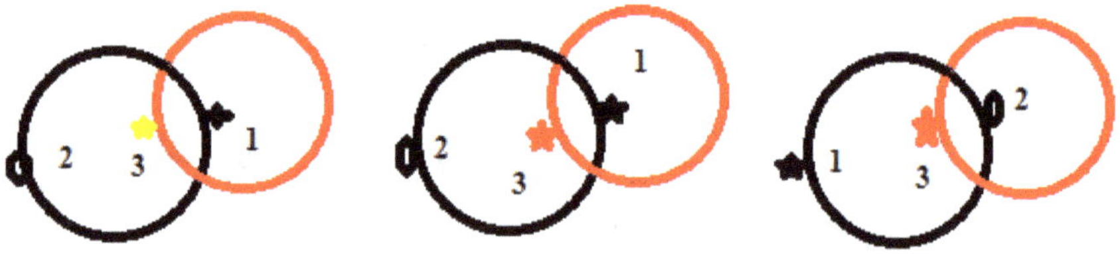

Ground state   **Site 3 switched off**   **Site 2 switched on**

Figure 3.5. Movements of rings between two sites by stimulus.

but in a distinguishable manner then the movement of rings can be controlled by varying stimuli.

Consider two interlocked rings having interacting sites at 1, 2 and 3 as shown in Figure 3.5. Considering the interaction between 1 and 3 are preferable, then the two rings will adopt a position as shown in Figure 3.5. Once the interaction of site 3 is switched by a stimulus, the option will force for interactions between site 2 and 3. Thus, repeated motion to the two ends will take place in a to and fro manner. When a moving ring of catenane has two different recognition sites, both sites can be turned off or on with different stimuli. Due to such movements, catenanes are comparable as molecular equivalents to ball, socket and universal joints used in macroscopic machines. Catenanes composed of three interlocked macrocycles show control rotation of molecular rings. Rotation movement path for one ring over another ring of the same catenane is circular, such rotational motion is equivalent to a linear movement restrained by cyclic geometry of the ring.

Catenane comprises of multiple rings, and the interlocking ring is a challenge during synthesis. Generally affinity of a particular portion of a ring to another component of another ring to be constructed is helpful in construction of catenanes. For example, ring **3.5** of Figure 3.6 has affinity for the cation **3.5a**. These two units can be brought together to contact to form template for synthesis of interlocked system **3.5c**. Upon reaction of the combination of **3.5** and **3.5b** with 1,4-dibromomethylbenzene forms catenane **3.5c**. Catenane **3.5c** has two non-equivalent interlocked rings.

In another methodology, metal template mediated reactions are used for preparation of catenane. One such example is illustrated in Figure 3.7. In this case the copper (I) complex provides a template which is suitable

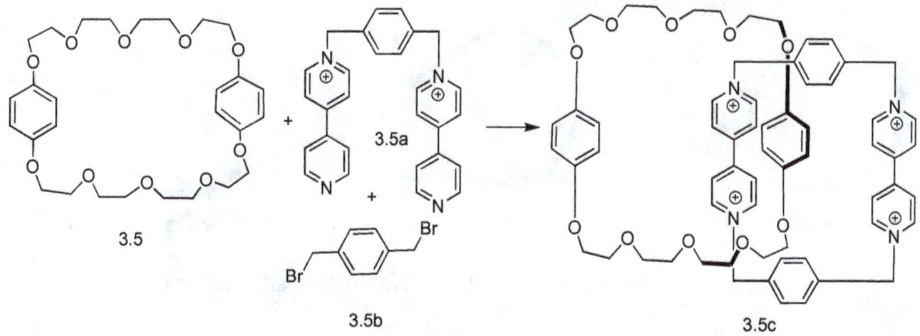

Figure 3.6. Synthesis of a catenane.

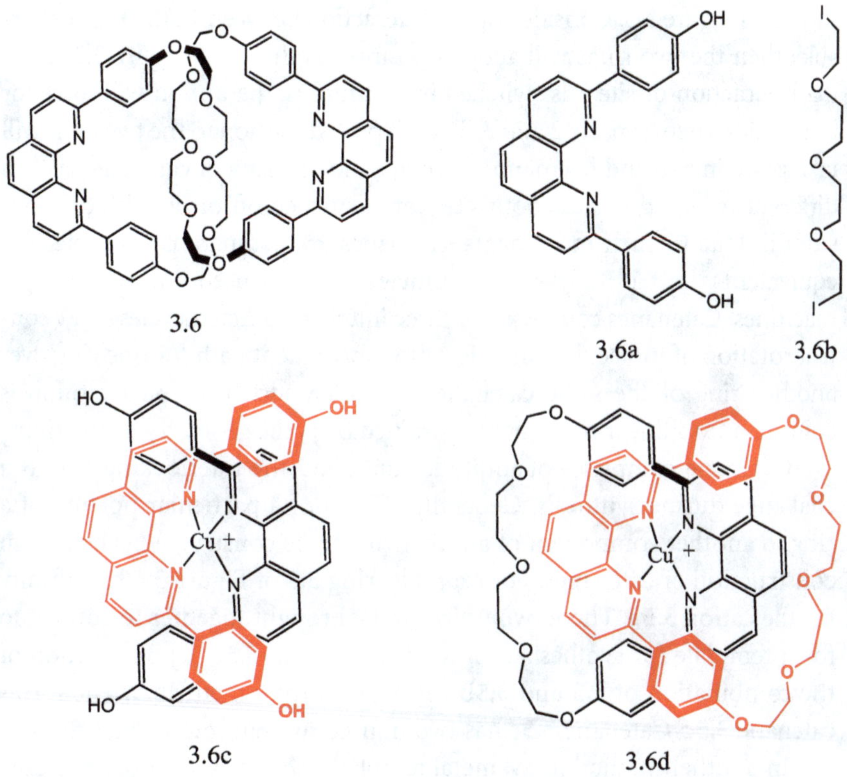

Figure 3.7. Copper(I) template directed synthesis of catenane.[3]

for formation of interlocked structure. A flexible ligand having two 1,10-phenanthroline and two functionalisable phenol groups are brought together to form a complex **3.6c**, this complex has a puckered structure, and it has appropriate geometry to undergo condensation with compound **3.6b** to form complex of the catenane **3.6d**. Copper ion from **3.6d** can be easily removed by treating it with cyanide ion giving the free catenane **3.6**.

Coordination ability of a metal ion is utilised for rotation of catenane rings. Copper(I) ion in coordination complexes prefers to adopt tetrahedral geometry by having suitably placed ligands. On the other hand, a copper(II) ion prefers penta or hexa-coordination. An example where advantages of just binding energy differences are taken to cause rotation within a catenane is described in Figure 3.8. Macrocyclic ring **3.7** has a 1,10-phenanthroline unit and a tripyridine unit at two independent positions. Three different types of coordination environments are

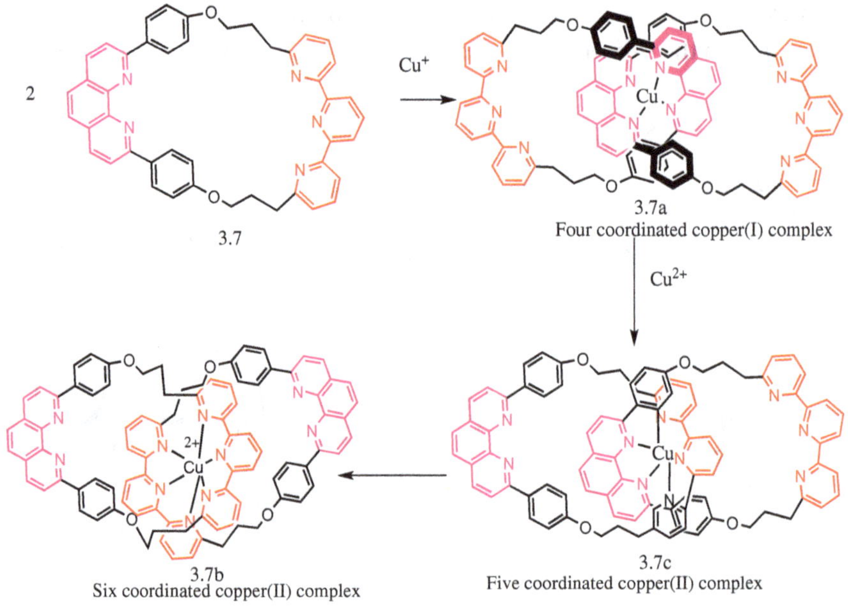

Figure 3.8. Copper ion mediated movement among the rings of a catenane.[3]

created among interlocked rings by bringing two ends of two rings. One is by bringing two 1,10-phenanthroline units together to preferentially coordinate to copper(I) ion. If the copper(I) ion is oxidised it adopts penta-coordinated state by shifting position of the ring and with time it slowly forms six-coordinated copper complex (Figure 3.8). Thus an initial tetra-coordinated state will transform to penta and hexa-coordination with copper ion upon oxidation of copper (I) to copper (II). On sequential oxidation and reduction of the copper ion will make rotation of the ring, which makes a net full rotation and process is a statistical distribution of oscillation and rotation. Oscillation involves the reorganization to penta-coordinated state to hexa-coordinated state, whereas rotation is the overall process.

Catenane can be constructed over a complex metal template or an organic template. Macrocyclic rings constructed on a porphyrin ring or porphyrin ring containing complex are used to as one ring to interlock another ring. One such combination is shown in Figure 3.9 is a zinc por-phyrin complex having an interlocked ring with another ring (**3.8c**). In this example of catenane rotation of hydroquinone part does not take place within the cationic ring **3.8a** due to high rotational barrier. However,

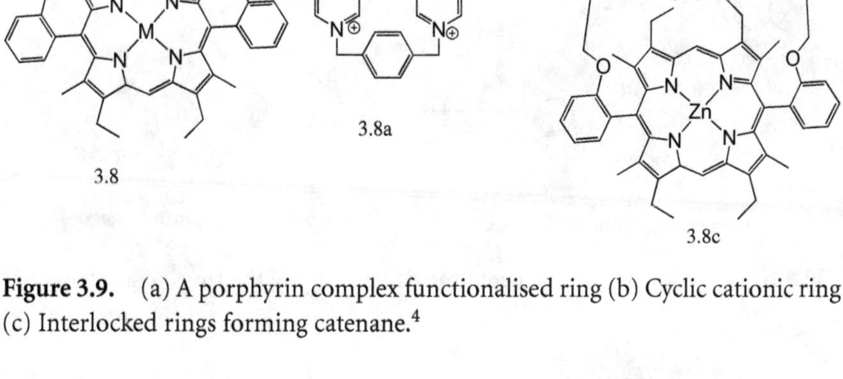

Figure 3.9.   (a) A porphyrin complex functionalised ring (b) Cyclic cationic ring (c) Interlocked rings forming catenane.[4]

in absence of metal ion, protonation or deprotonation of porphyrin ring displaces the position of the hydroquinone part with respect to porphyrin ring causing a mechanical motion.

3.3 Rotaxanes

Interlocking of compounds having ring-like structure with another molecule with dumb-bell shaped structure makes a class of compounds called rotaxanes. Rotaxane are those molecules which have macrocycles held by a portion of a dumb-bell-like component. This class of interlocked molecule have dumb-bell shaped compound holding one or more ring. Generally the two ends of dumb-bell type of structures possess two bulky groups at the ends. These bulky groups help to retain these rings by acting as barrier so that ring does not easily get dislodged from such system (Figure 3.10). When more than one cyclic hydrophobic component is present, such cyclic components generally maintain optimum distances among the rings. This is due to repulsion between the rings. But their

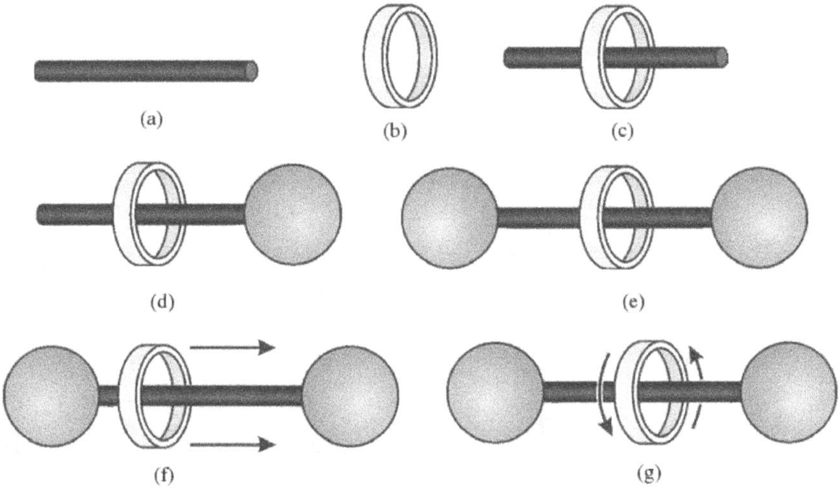

Figure 3.10. (a) and (b) are the principle components of rotaxane; Examples shown in (c)-(e) are different components of a rotaxane and (f) and (g) are two types of motion in a rotaxane.

locations in rotaxane are also decided by two other factors, one is the interaction with a specific site on the rod and /or interactions between the cyclic components. There is also another kind of interlocked system that may be formed by combining ring-like element and a thread-like element, they are called pseudorotaxane. This class of compounds can be threaded or dethreaded onto the ring more easily by various stimuli. There are many examples of cyclic units made of polyether or aza-crown. These two classes of compounds possess remarkable guest recognition properties hence their presence in rotaxanes have advantages.

Macrocycles component of a rotaxane may have different binding sites with slight variations in energies, hence can be switched reversibly between such sites with the aid of a suitable stimulus. Such stimuli can be chemical, electrochemical, or photochemical. There are several classical ways to change stereo-electronic properties of a binding site. It can be done by protonation or deprotonation, or by electron transfer process such as oxidation or reduction. Such alteration of binding affinity at a particular binding site either of the rod or the cyclic component of a rotaxane can be utilized for movement of a ring across the rod-like molecule. The stereo-electronic property of a recognition site within a macrocycle can also be varied and can be utilized to move a ring forward or backward in a rotaxane as illustrated in schematic Figure 3.11. Among

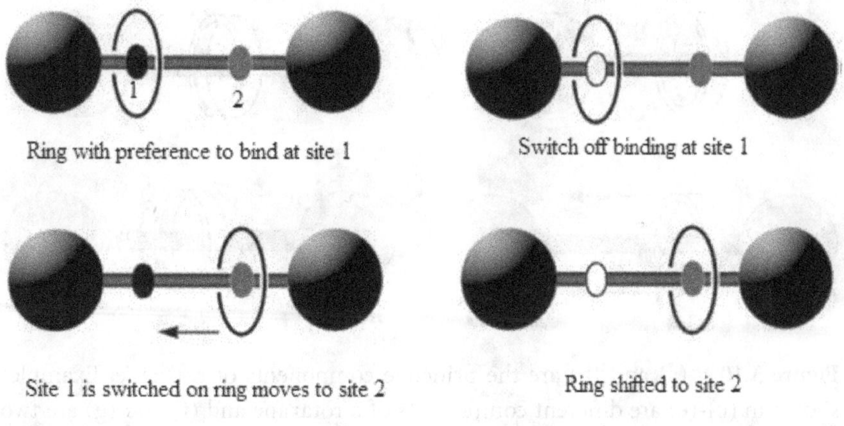

Ring with preference to bind at site 1   Switch off binding at site 1

Site 1 is switched on ring moves to site 2   Ring shifted to site 2

Figure 3.11. A scheme to describe forward and backward movement by stimuli.

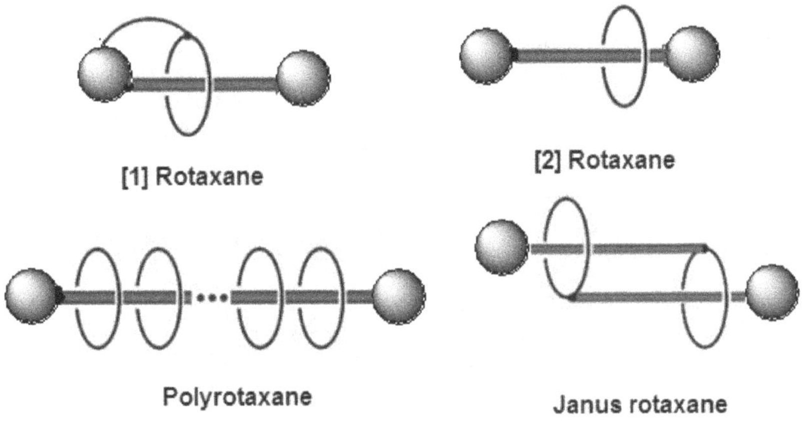

Figure 3.12. Schematic representation of different types of rotaxanes.

two binding sites on the axle of rotaxane one may have higher affinity (site 1 in Figure 3.11) over the other site. In such instance, the strength of steric force and applied force guides rotation of macrocycle to adopt a stable conformation.

Rotaxanes are classified as [1]rotaxane, [2]rotaxane, Janus rotaxane and polyrotaxane (Figure 3.12). In [1]rotaxane, the ring component is attached to one of the terminal, whereas in [2]rotaxane ring is freely held within a rod of dumb-bell shaped geometry. Polyrotaxanes have more than one ring held by the rod part of the rotaxane. On the other hand, Janus rotaxane have interlocked structures. There is another class of rotaxane called pueudo-rotaxane which is formed by a rod and a ring-like structure without dumb-bell structure.

3.4 Thermodynamics of Interlocked Systems

One widely used host system for understanding design of host-guest complex and rotaxane is tetra-cationic cyclophane (**3.9**) illustrated in Figure 3.13, it can act as host for wide ranges of guest molecules (**3.10–3.17**) such as shown in Figure 3.13. Study on the thermodynamic parameters and the influence of different factors provides the fundamental basis for the construction of machines derived from interlocked systems.

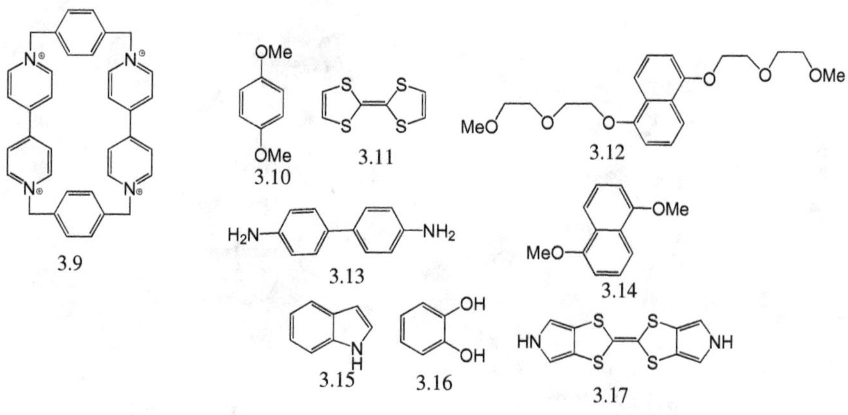

Figure 3.13. A tetra-cationic electron deficient ring and series of donor molecules.

Dumb-bell shaped guests containing a π-electron-donating unit inter-acts with π-electron-deficient tetracationic cyclophane forming [2] rotaxanes.

On the other hand, constructing dumb-bell shaped guest and by introducing stoppers at two ends of such guests provides scope to make interlocked systems. Rotaxanes can be formed by host-guest interactions by two paths. One path involves clipping of guest into host, by statisti-cally probabilistic insertion of the guest molecule into the cavity of the host as illustrated in Figure 3.14. Other option is to form a host-guest complex by inserting a rod-like guest to void of host and two ends of the guest molecule are functionalised by bulky functional groups to serve as stoppers.

There are two important aspects relevant to these two methodologies in constructions of rotaxanes. First point is the nature and strength of host guest interactions between the ring and the rod. Second point is the energy minimum in the potential energy surface for such process to cross the activation barrier to form a rotaxane (Figure 3.15). Thus, first part relates to host guest interactions and second part is the formation and stability of a rotaxane. Association constant K_a for the host-guest complex is defined as the concentration of the complex, divided by the product of concentration of the free components.

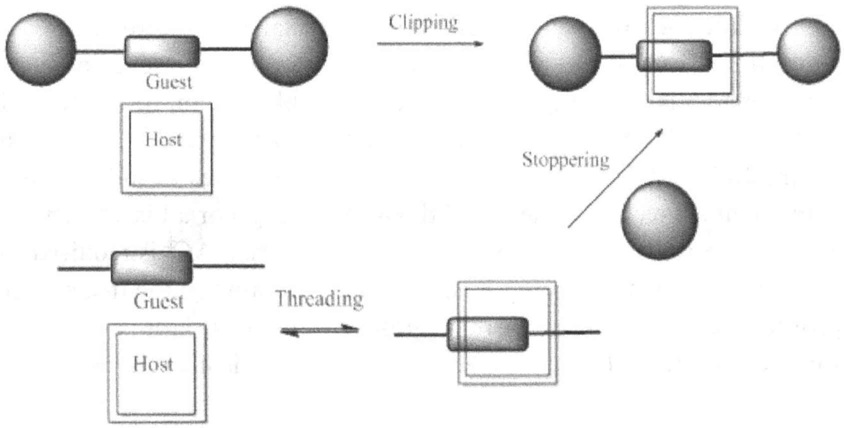

Figure 3.14. Two different paths for construction of rotaxanes.

In general association constant K_a at particular temperature of a host guest complex is exponentially dependent on the difference in free energy between the separate components and the complex.

$$K_a = [\text{Complex}] / [\text{Host}][\text{Guest}] = e^{-\Delta Ga/RT}$$

$$\Delta G = \Delta H - T\Delta S$$

$$\Delta G = - RT \ln K = \Delta H - T\Delta S$$

$$R = \text{Universal Gas Constant } (8.3144 \text{ Jmol}^{-1} \text{ K}^{-1}).$$

Rate constant of host-guest complex is dependent on the free energy barrier, hence free energy barrier associated with each state is responsible to dictate the thermodynamic and kinetic properties of the mechanically interlocked molecules.

$$k = e^{-\Delta G\ddagger/RT}$$

$$\Delta G^{\ddagger} = - RT \ln (hk / k_B T) = \Delta H^{\ddagger} - T\Delta S^{\ddagger}$$

$$h = \text{Planck constant } (6.262 \times 10^{-34} \text{ J s})$$

$$k_B = \text{Boltzmann constant } (1.381 \times 10^{-23} \text{ m}^2 \text{ kg s}^{-2} \text{ K}^{-1}).$$

For interlocked systems, such as in [2]rotaxanes, rings are held over a dumb-bell shaped rod bearing large bulky stoppers. Presence of stoppers provides large energy barriers making stable system. Interaction of ring with specific binding site has a potential energy profile as shown in Figure 3.15. In this diagram bottom of the well relates to independent components. Free energy associated with it is ΔG_a. For a bistable rotaxane, double-well surface with an intervening barrier, $\Delta G^{\ddagger}$ is required to describe the potential energy surface. Each well found in the free energy profile shown in Figure 3.16 represents one translational isomer. Among the translational isomers more favoured isomer is called the

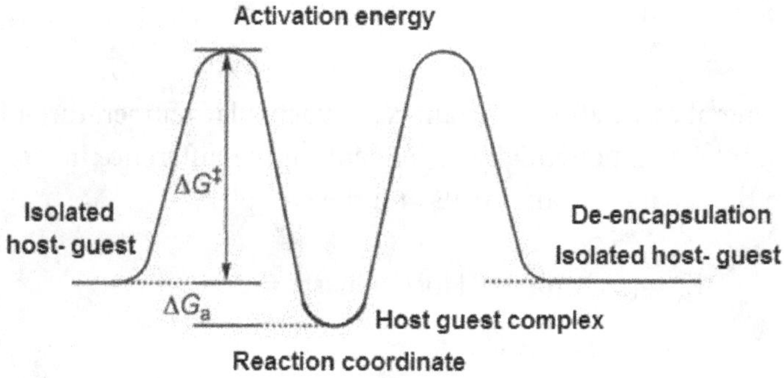

Figure 3.15.   Potential energy changes during host-guest complex formation.

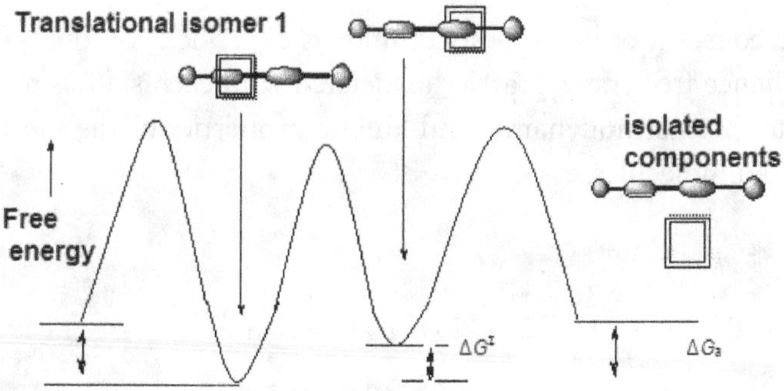

Figure 3.16.   Potential energy diagram of bistable [2]rotaxane showing two translational isomers.

ground-state co-conformation (GSCC) and the less favoured one is called as metastable state co-conformation (MSCC). Two recognition sites and co-conformations lead to bistability. Free energy difference between the potential wells is a measure to distinguish affinities of the ring for selective site and it also relates host-guest complexation. The equilibrium constant, K_{eq}, for bistable rotaxane is defined as ratio between populations of one translational isomer over the other, N_{GSCC}/N_{MSCC}. $K_{eq} = N_{GSCC} / N_{MSCC}$. A [2]rotaxane have two identical recognition sites is called degenerate rotaxane which is encircled by one cyclophane or by suitable ring-like structures. In such cases the potential potential energy profile is symmetric and has $\Delta G = 0$. At this condition movement of the ring averages out and equilibrium is reached. For this purpose $\Delta G^{\ddagger}$ is a very important parameter to effect a shuttle-like motion in interlocked system and this free energy difference between the translational isomers determines the rate of shuttling between two selective sites.

Donor strength associated with ring of rotaxane play a crucial role in deciding the association constants. For example, tetrathiafulvalene has E_{ox} = +0.34V is a strong donor has an association constant K_a = 6900 M^{-1} with tetracationic cyclophane **3.9**. On the other hand, benzidine (E_{ox} = +0.59V), 1,5-dimethoxynaphthalene (E_{ox} = +1.11V), and 1,4-dimethoxybenzene (E = +1.31V), have respective K_a for complex with cyclophane **3.9** are 1044.0 M^{-1}, 440.0 M^{-1} and 18.0 M^{-1} respectively. Inclusion complex of 1,4-dihydroxybenzene and 2,4-dinitrophenol with cyclophane **3.9** have C–H$\cdots\pi$ interactions, with hydrogen bond donor acceptor distance ~2.8°A. They also show π–π separation ~3.8°A. Another important interaction found in the host guest complexes is C–H$\cdots$O the α-protons of the bipyridinium rings of the cyclophane with oxygen atoms of Diethyleneglycol chain of host like **3.13**. In such cases hydrogen bond donor-acceptor distances are in the range of ~3.2–3.3°A. The size of a π surface involved as donor has contribution to binding constant. For example, a phenyl ring and naphthalene rings differ in sizes so as the π–π stacking interactions. Interactions of bipyridinium ions of tetracationic cyclophane host with a π-electron rich guest lead to a more stable complex. Solvent also contributes to binding between hosts and guests. For example a solvent interacting with higher binding affinities are observed by using more polar solvents. Complexation of the electron donor solvent with the π-electron of cyclophane decreases the area of the aromatic surfaces. In general a polar solvent has low molecular polarizabilities and large cohesive forces which facilitate complex formation. When two sites at different locations of the dumb-bell shaped rod for binding to ring exists than the higher association constant K_a for complexation between model guests and host ring lowers bottom of the ground state potential well, such situation will have higher free energy difference than the other state. Consequently, to have selectivity larger barriers will show stronger molecular recognitions.

Spacers present in a dumb-bell serving as axle of rotaxane has relatively small contribution to the free energy barriers $\Delta G^{\ddagger}$ for a molecular shuttle. Such energy differences are about 0.5 kcal mol^{-1}. But other thermodynamic parameters such as enthalpy change ($\Delta H^{\ddagger}$) and entropy changes ($\Delta S^{\ddagger}$) are significantly affected by spacers. For example, a flexible tether may adopt multiple orientations and may undergo conformational changes. These aspects are not possible with rigid tethers. Conformational aspects have great consequences in performances of molecular machines. Entropy and enthalpy aspects are decisive to narrate merit and performance. Thus, enthalpy and entropy changes in a wide span of temperature provide a correct picture on operational limits and help to differentiate functions for a particular application. A fast shuttling action is achieved by using a spacer having small negative entropy and enthalpy changes.

3.5 Shuttling Motion in Rotaxanes

One of the earliest rotaxane described by Stoddart group is comprised of cation of cyclophane ring **3.9** incorporated in a linear polyether **3.18a**. The end groups of the rods are tri-isoproylsilane connected by Si-O bond to a polyether which is connected to hydroquinone and glycol ethers as shown in Figure 3.17. When **3.9** is threaded to this rod-like structure it tends to reside at position where the phenylene ether is located. Due to a low activation barrier the ring can exchange positions solution. The equilibrium can be studied in acetone-d$_6$ solution with the help of nuclear magnetic resonance spectroscopy. At ambient temperature the rotaxane in acetone-d$_6$ show broad signals for the OCH$_2$ groups appearing in the

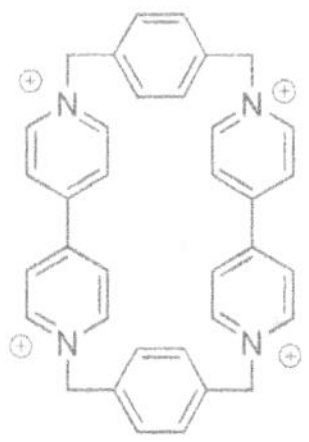

3.18a

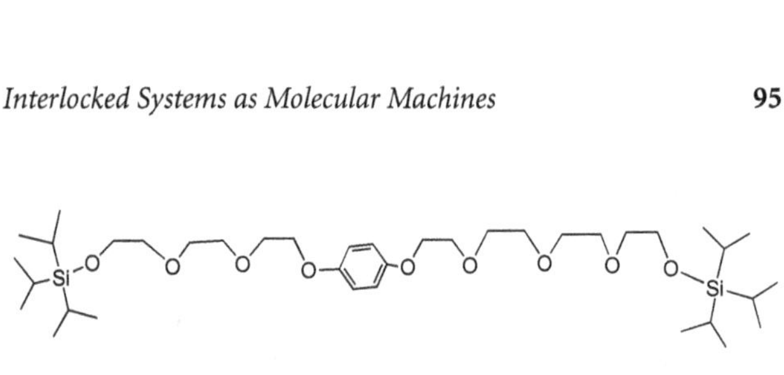

3.9

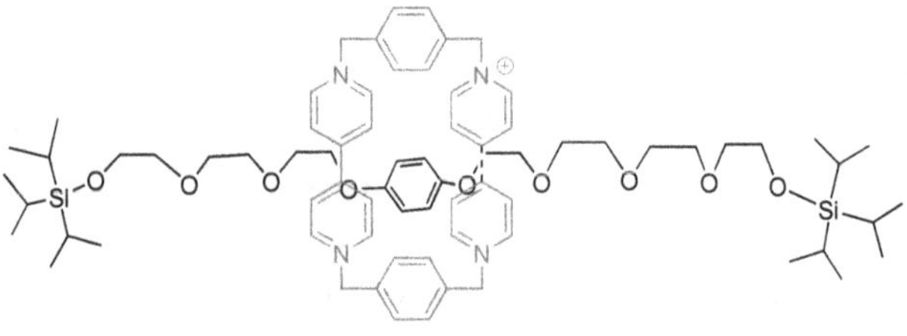

3.18

Figure 3.17.   A rotaxane with phenylene ether and glycol ether as sites for inter-action with cationic ring.

region of 3.65–3.96 ppm and the signals for the hydroquinol ring pro-tons are very broad and are merged in with the base line.[5]

On cooling the solution the peaks are resolved and the hydroquinone peaks becomes distinct. The rotaxane when dissolved in dimethylsulphox-ide and spectra is recorded at warm condition an AA′BB′ pattern having centered at 5.16 ppm arises from hydroquinone part. From such a study the energy difference between the two states are found to be 13Kcal/mole. Most important aspect to design a simple molecular machine is to have an efficient switching process and bistability. Rotaxanes and catenanes are interlocked system which provides basis for degenerate (molecular shut-tles) and switchable bistable systems that are influenced by environment. Study on cyclophanes have provided breakthrough in understanding fundamentals of molecular machines based on interlocked systems. Cyclophane are good substrate to interact with guests through π–stacking and also by charge-transfer interactions. Some of them are useful and with to interact with guest molecules through C–H⋯O interactions. Furthermore cyclophanes possess rigid cavity which are useful to hold guest molecules in reversible or irreversible manner.

3.6 Switching in Rotaxanes

First switchable [2]rotaxane was reported with a dumb-bell axis having two π-electron-rich donor units one was derived from benzidine and other was from 4,4'-biphenol ether illustrated in Figure 3.18. The tetracationic cyclophane ring of this [2]rotaxane shuttles showed two translational isomers in 84 : 16 ratio in deuterated acetonitrile at −4°C. The cyclophane tetra-cation was preferentially located over the benzidine site. This was due to presence of relatively higher amounts of π-electron at this site due to benzidine over the π-electron density at the phenolate ether site. Addition of excess amounts of deuterated trifluoroacetic acid to the solution of rotaxane caused a movement of the cationic cyclophane unit from the benzidine site to the ethereal site. This is due to protonation of the amine groups to generate ammonium ions which repel the cationic charge of the cyclophane ring. On the other hand, addition of deuterated pyridine reinstated the ring to the original position or in other word the original state could be reverted by neutralisation with pyridine.

Rod-like structure **3.20** has two biphenyl units located in different environment with two bulky tri-ispropylsilyl groups at two ends. A tetracationic bipyridinium ion derivative with cyclic structure acts as a

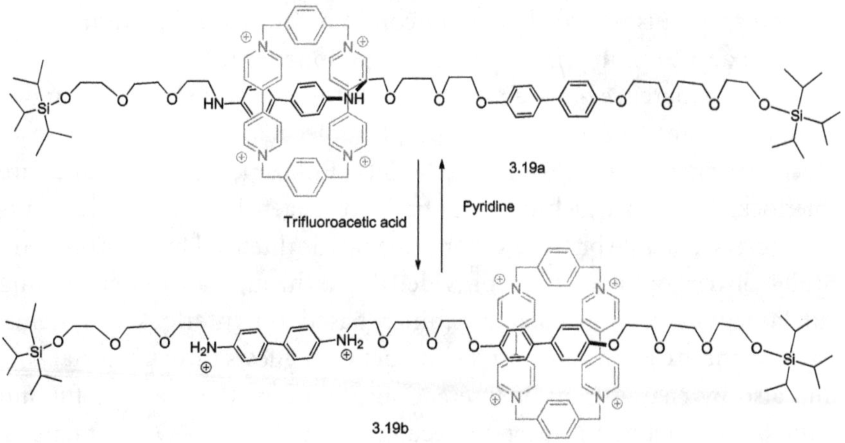

Figure 3.18. Shuttling action of a tetracationic cyclophane ring over a dumb-bell shaped molecule of a rotaxane.[6]

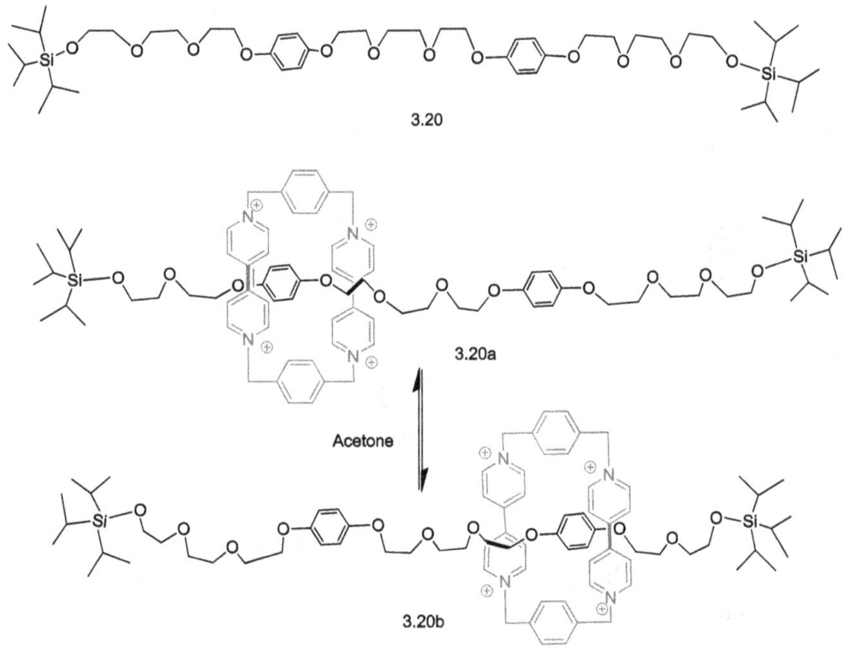

Figure 3.19. (a) A molecule with dumb-bell shape, (b) Equilibrium of rotaxane due to equal probabilty of finding the ring at two equivalent position of the rod.[7]

thread at either of the ends with equal probability due to having phenylether groups interacting with the bipyridinium cations on the ring (Figure 3.19). Energy difference between the two conformation isomers is 13 Kcal/mol.

Calix(4)arenes are phenolic derivatives having cyclic structures. Mechanised motion in a ring-like structure involving calix(4)arene derivative with such as **3.23** and a rod-like molecules such as long-chain attched to bipyridine (**3.22a** or **3.22b**) can be carried out. Such a mechanised motion can be triggered by energy obtained from another photochemical equilbrium operative simultaneously. In this case insertion and release can be controlled by opertaing photochemical ring opening between the compounds **3.23a** and **3.23b** (Figure 3.20). The linear structure having phenolic group in **3.23a** cyclises upon irradiation to form **3.23b** and this imbibes competition of insertion of bipyridinium cation host with **3.23a**

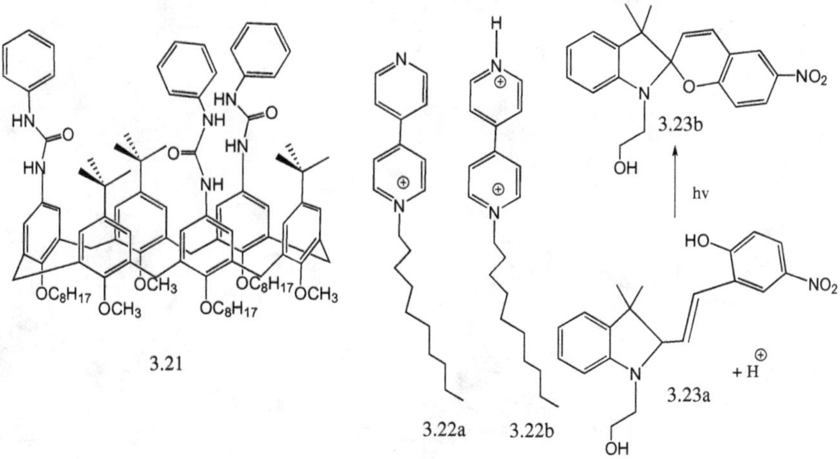

Figure 3.20. A calix-6-arene derivative (**3.21**) as ring and N-substituted proto-nated bipyridine derivatives (**3.22a** and **3.22b**) as axle components of a rotaxane which is mechanised by using reversible photo-chemical equilibrium for energy source from conversion of **3.23a** to **3.23b**.

to the void of calix-host, leading to insertion and elimination guest at the cost of the photochemical reaction.[8]

Different components of rotaxanes can be assembled by template-directed synthesis such as threading, clipping, and slippage.[9] In Figure 3.21 the rod-like structure **3.24** has two biphenyl unit in different environment with two bulky R groups at two ends. A ring **3.25** a can be threaded at either ends with unequal probability due to biphenyl groups which can interact differently with bipyridinium cationic parts on the ring.

Most conventional way of supplying energy to a rotaxane to show properties of a molecular machine by using some chemical reactions that release energy. For example combination of pyridine derivative and the ring **3.26** forms rotaxane with a molecule **3.27,** which has two bulky stop-pers at two ends. In rotaxane **3.28a** the ring is stabilized by diaminomethylene pyridine unit through N-H···N hydrogen bonds. Once this form of the rotaxane is treated with a strong acid like trifluorometh-ane sulphonic acid, nitrogen atoms of NH and pyridine are protonated (Figure 3.22).

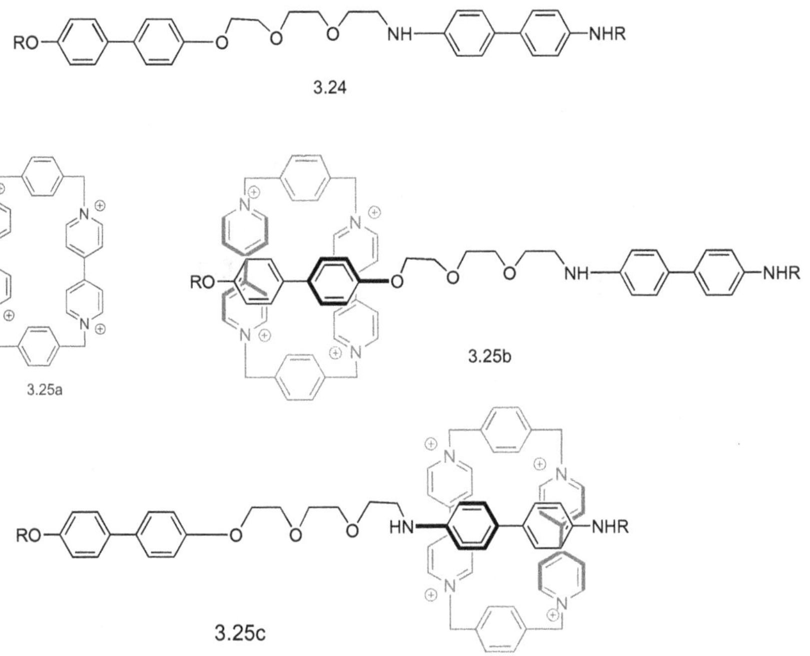

3.24

3.25a

3.25b

3.25c

Figure 3.21.   Threading of rings at two non-equivalent ends.

Protonated sites can hydrogen bond with the anionic counterpart. Hence, to accommodate trifluoromethane sulphonate anion in the scaffold, the sites and mode of hydrogen bonds change and it forms another rotaxane **3.28b**. Upon protonation pyridine containing site turns to pyridinium cation and it becomes hydrogen bond donor site which was originally a hydrogen bond acceptor site. This causes a movement of the ring. The protontated form when treated with sodium hydroxide, original hydrogen bonded form is recovered. This movement caused by hydrogen bonds results in folding of the rotaxane and also provides rigidity to the new state of the system.

There are examples where weak interactions other than hydrogen bonds play roles in the movement of ring in a rotaxane. One such example is shown in Figure 3.23 is based on π-electron donor-acceptor interactions. In this case [2]rotaxane has macrocycle **3.29** preferentially interacting with 1,4,5,8-naphthalenetetracarboxylate diimide unit. Naphthalenediimide has

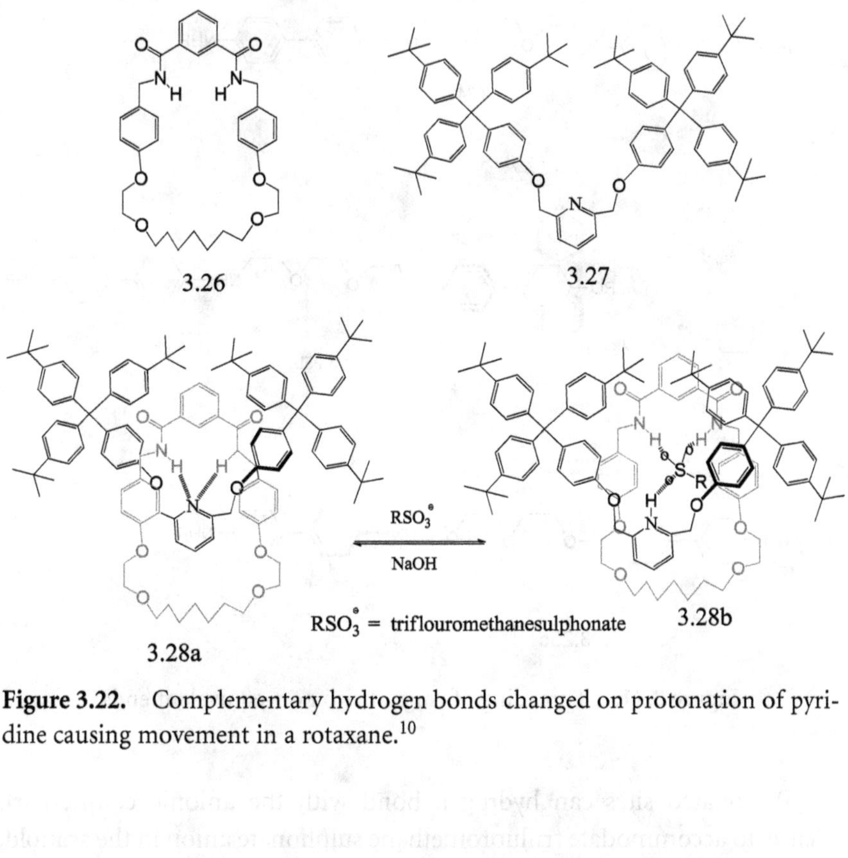

Figure 3.22. Complementary hydrogen bonds changed on protonation of pyridine causing movement in a rotaxane.[10]

better electron withdrawing ability than pyromellitic diimide unit. When lithium perchlorate is added to a solution of this rotaxane **3.31a**, strong interactions of lithium ion with oxygen atoms in the polyether loops of the macrocycle takes place. Such interactions help to stabilize the ring over the region where pyromellitic diimide unit on the axle of the rotaxane is located. Thus, the ring moves from one position to another position. On the other hand, upon addition of [12]crown-4 to this state removes the lithium ion by selective binding of lithium ion with the externally added crown ether. The crown ether has higher affinity for the lithium ion to remove it from the rotaxane. The abstraction of lithium ion causes the ring to move to the original stable state. Thus, addition of lithium perchlorate causes a forward motion whereas addition of [12]crown-4 takes back the ring to original position.

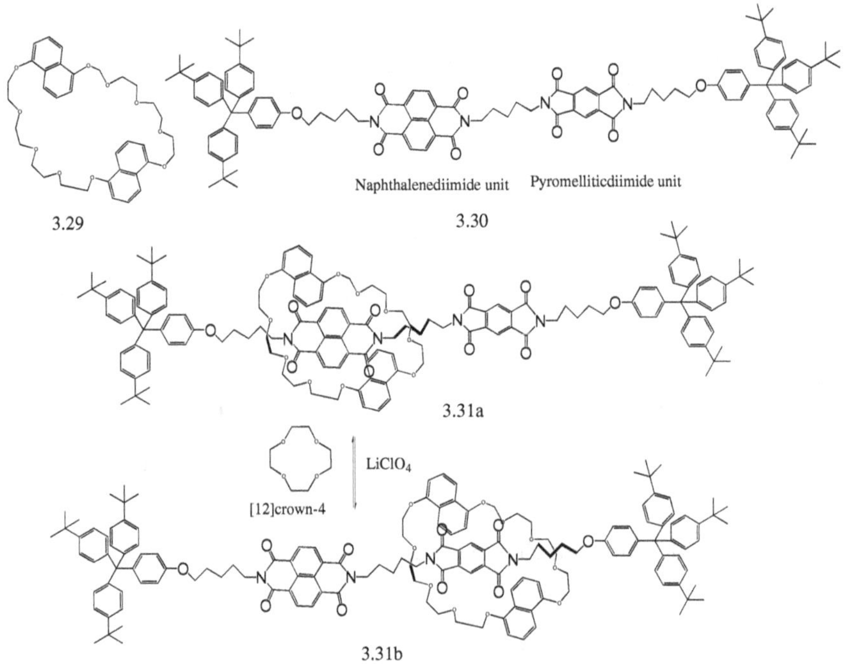

Naphthalenediimide unit   Pyromelliticdiimide unit

3.29   3.30

3.31a

[12]crown-4 | LiClO$_4$

3.31b

Figure 3.23.   Utilisation of electronic effects to shuttle a ring of a rotaxane.[11]

Thermodynamic stabilty of different self-assemblies can be a guiding factor to cause movement in a rotaxane. Such process happens in roraxanes which can fold or unfold rod-like structure to make bend or to provide a linear geometry. In the example shown in Figure 3.24, the rotaxene **3.34a** adpots a folded structure by utilising complementary hydrogen bonds of amide bonds of the rod with 2,6-pyridinedicarboxydiamide derivative. On heating the molecule gets stretched to form new set of hydrogen bonds. Thus, by this process an orderly structure **3.34b** of rotaxane is formed from a less orderly state, which is thus termed as entropy guided transformation. Thus in this rotaxane state **3.34a** at 258 K is stable which on heating to 308 K changes to state **3.34b** and vice versa. This leads to contraction and expansion of the rotaxane.

Change in polarity of medium causes movement of rings in a rotaxane **3.35a**. Similar movement is easily executed in certain rotaxanes

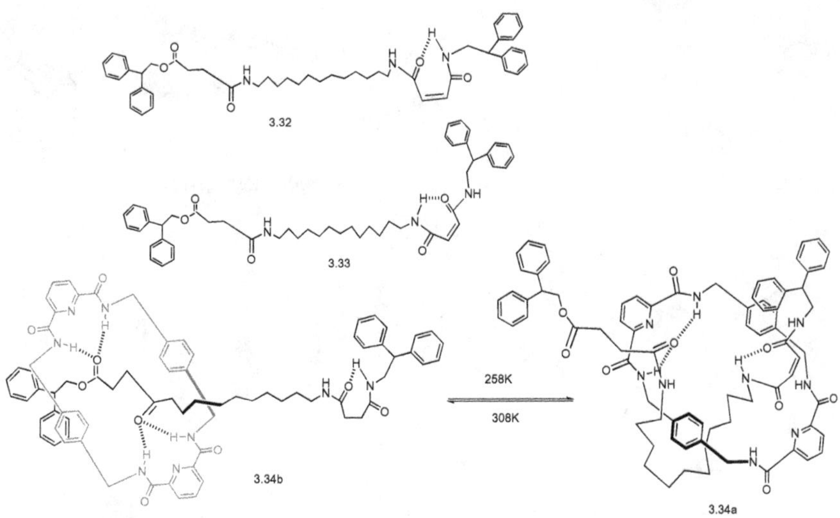

Figure 3.24. Movement of ring in rotaxane takes place as well as the intramolecular hydrogen bonding state changes.[12]

having hydrophobic and hydrophilic part. Example shown in Figure 3.25 is a rotaxane where there is amide bonds next to a bulky C_{60} unit. It has also a hydrophobic long chain aliphatic hydrocarbon unit to connect a bulky ditertiarybutyl phenoxy as stopper. A 2,6-dicraboxyhenzene based molecule **3.36** having ring-like structure can be bound to the amide bond so that it takes up a position next to the C_{60} unit. The ring unit moves from hydrophilic part once a more polar solvent favoring hydrogen bond with the substrate such as dimethylsulphoxide. This is due to the ability of the solvent to interact with the amide bonds of the rod-like structure **3.37**.

Molecular switch as logic GATE and memory devices are constructed from [2]rotaxane anchored to a surface through a polymeric unit. For example rotaxane **3.38** has a fluorophore at one end a functionalisable polymeric unit at another end, it has also amide bonds for hydrogen bonds (Figure 3.26). A part of the rod is having hydrophobic aliphatic hydrocarbon skeleton namely a (-CH_2)$_9$- unit. In this rotaxane hydrogen bonds controls the shuttling motion of the ring **3.40** upon formation and

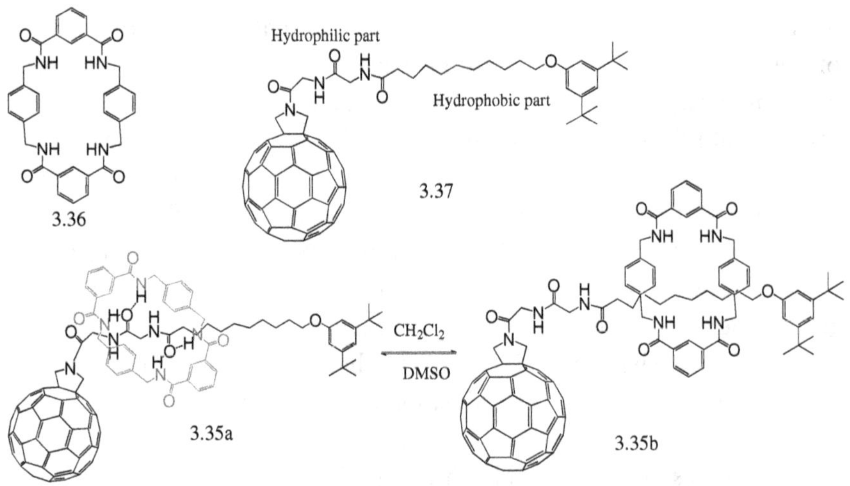

Figure 3.25. C_{60} functionalised rotaxane which has two sites for a ring to move.[13]

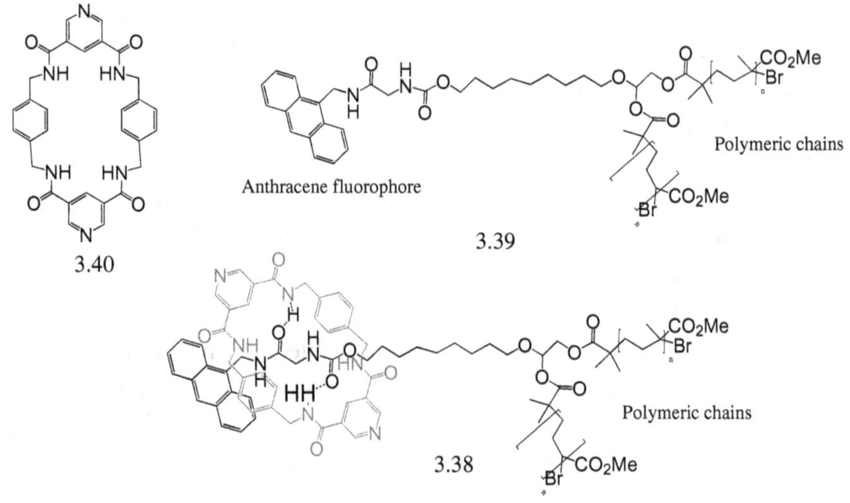

Figure 3.26. Modulation of fluorescence by shuttling motion of ring in a rotaxane.[14]

disruption of hydrogen bonds are by a polar solvent. Changes of positions of the ring with respect to the dumb-bell create switches with fluorescent on/off intensity ratio. The process occurs due to change of photo-electron transfer path during fluorescence between the different components. Such changes are observed in solid state [2]rotaxane having one end attached to a polymer. A film of such polymer having rotaxane as a functional part, movement of the macrocyclic from the amide bound state to the alkyl chain is induced by a DMSO vapor. This rotaxane has a fluorescent anthracene group. The fluorescence of the rotaxane occurring due to anthracene fluorophore is quenched in acid vapor. This happens due to protonation of pyridine unit on the ring by acid vapor, and the fluorescence is recovered neutralizing the protonated and during such deprotonation pyridine macrocycle shuttles to the portion having alkyl chain.

Due to a ring-like component with a rod-like component in rotaxane at least two molecular motions are associated with such molecules. One is (i) translation motion, which describes shuttling of the ring along the axle and other is the (ii) rotation motion of the macrocyclic ring around the axle. Thus, rotaxanes have the features to act as rotary and linear molecular motors. As far as molecular motors based on rotaxane are concerned in majority of cases they function as molecular shuttles. For this purpose the rotaxane utilises two or more distinguishable binding sites within the dumb-bell portion. Two identical sites these sites are called degenerate sites. As a result of equal probability of binding at these two sites results equilibration of the ring between two positions to make back and forth motion. However, when such sites are different they are called non-degenerate sites and relative free energy decides the movement of the ring component. In such situation the molecular ring resides at a particular place of the axle which is like a stable station called as state 0, but as soon as stimulus is applied this state gets switched off and it moves to another state by Brownian movement, which may be called as state 1. If state 0 is switched by an opposite stimulus, the original state is regained.

An imine group is easily hydrolysed to form carbonyl and such reaction is reversible under suitable condition. This reversible process is utilised in rotaxanes to covalently anchor a macrocyclic ring at a particular place. By hydrolysing imine group, carbonyl group is formed and ring-like structure gets modified. Accordingly the position of the ring in the rotaxane is changed. An example shown in Figure 3.27, the rod part of the

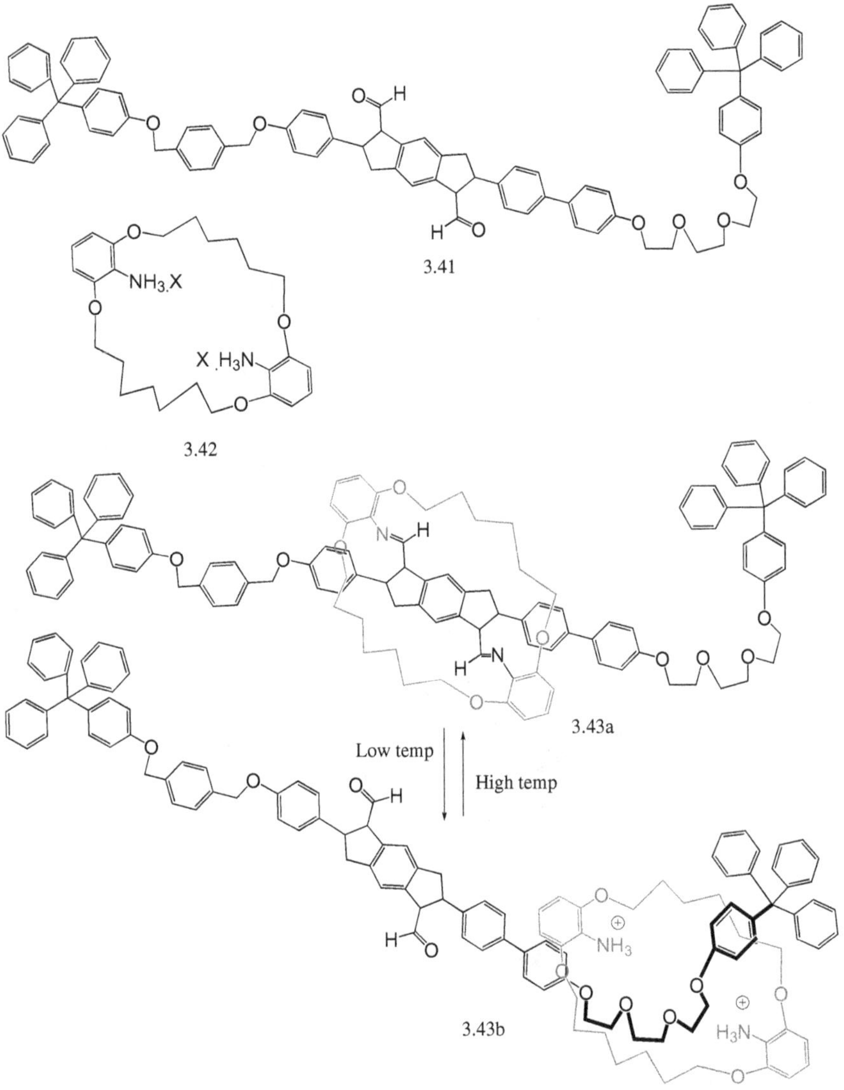

Figure 3.27. Hydrolysis of C=N bond to control shuttling motion of a ring in a rotaxane.[15]

rotaxane is a dialdehyde derivative. A macrocycle **3.42** is locked by imine bond formation to form a rotaxane like structure **3.43a**. On hydrolysis diammonium cation is formed and the as a result macrocyclic ring moves generate another state **3.43b** in which the ring is located at the part of the

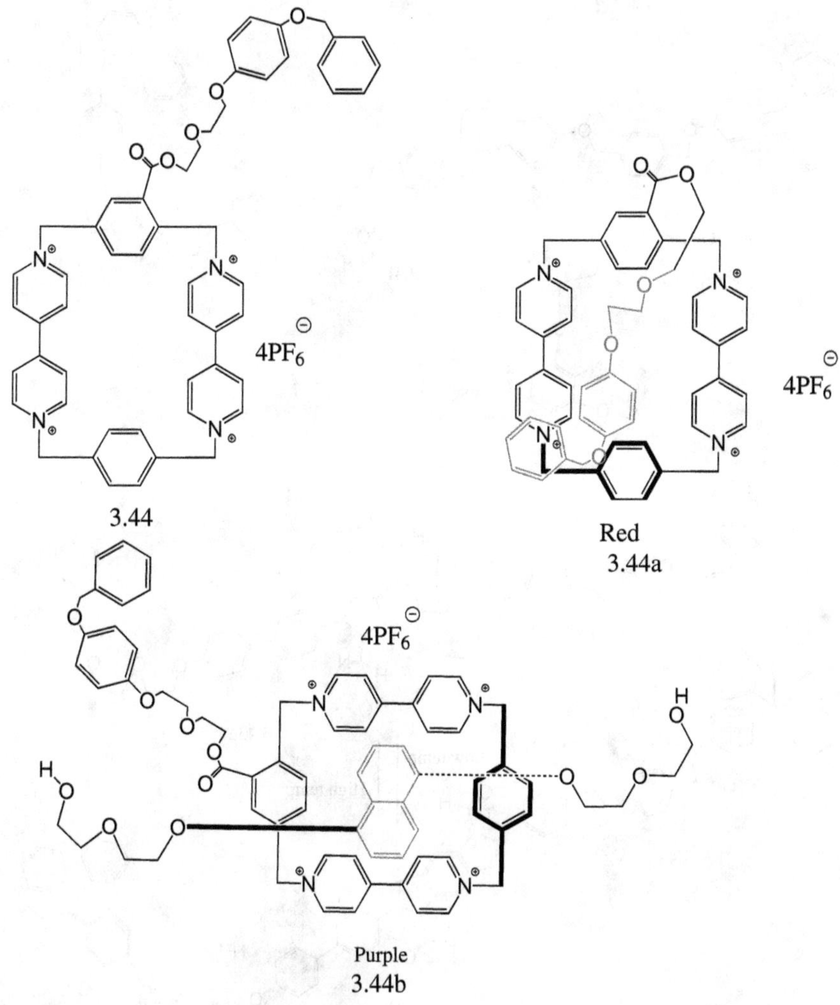

Figure 3.28.   Competitive binding to insert an arm into cavity of a macrocycle.[16]

axle containing polyether. This happens due to higher affinity of ammonium ion for polyether.

The example given in Figure 3.28 is a mechanised movement of a tether acting as an arm of a macrocycle **3.44**. The native state of this molecule **3.44a** has arm inserted into the cavity of the macrocycle. This inserted structure is stable due to π-interaction between the rings present in the macrocycle with the electron rich aromatic ring on the arm. The native folded form can be changed to a stretched geometry **3.44b** once it is treated with another polyether having a naphthalene unit. Naphthalene ring has higher affinity to bind with the ring forming π-interactions. Hence, the covalently linked arm of the macrocycle gets expelled out from the cavity of macrocyclic ring. This exit of the arm from the macrocyclic ring, facilitate insertion of the externally added naphthalene containing polyether molecule to form a new rotaxane **3.44b** and vice-versa.

Mechanical motion can also caused in interlocked systems possessing a flexible arm containing rotor like unit attached to a crown ether. The ammonium salt **3.45a** has a restrictively rotatable double bond (Figure 3.29). The olefinic unit having connected to two units comprising of three fused ring each has a chain which can be inserted to a cavity of crown ether located at another side of the molecule. The compound being an ammonium salt it has a structure in which flexible arm is inserted to the crown part. This is due to recognition ability of the crown ether towards ammonium cation. When the ammonium cation is neutralised to form amine by setting the hydrogen ion concentration of the medium to neutral condition the arm is released from the cavity of the macrocyclic ring. Such release and insertion by change of proton concentration causes mechanical movement of the molecule.

Designed rotaxane based ligands for metallo-organic frame work structures are useful in making complex machines. For example, a MOF constructed from tetra-carboxylic acid having rotaxane like structure as shown in Figure 3.30 has possibility for rotational and shuttling motion

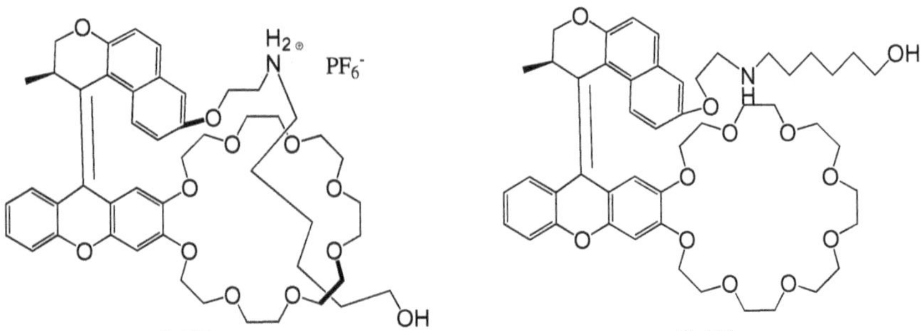

3.45a   3.45b

Figure 3.29. Movement of an arm attached to rotatable component.[17]

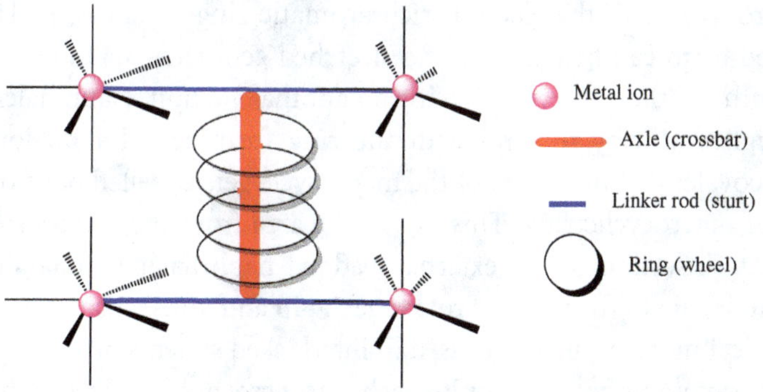

3.46

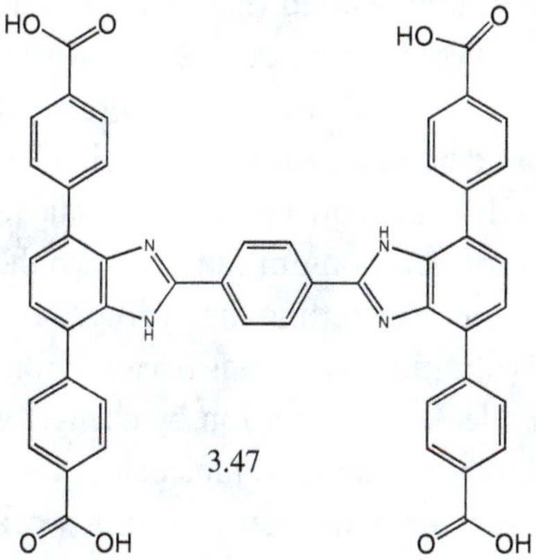

3.47

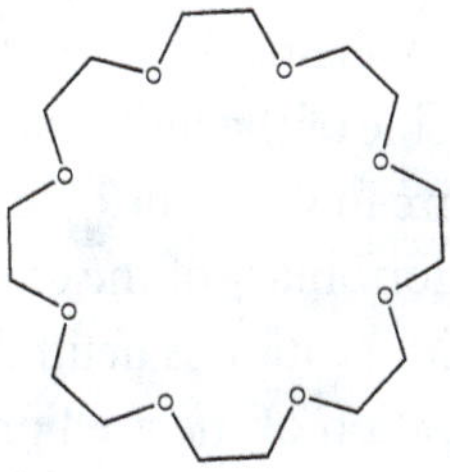

3.48

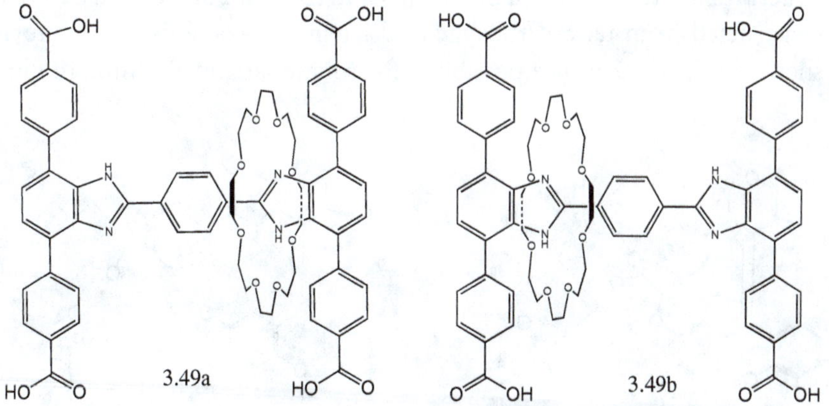

3.49a   3.49b

Movement of ring between two equivalent positions

Figure 3.30. Rotaxane based tetracarboxylate ligand showing movement of rings.[18]

on a rod-like structure. For example the tetracarboxylic acid shown in Figure 3.30 has a macrocylic ligand that can equilibrate at two positions due to presence of two equivalent imidazole units at two equivalent positions. Same compound on protonating and deprotonating at the tertiary nitrogen atom can cause rotational motion. The ligand has been utilised to prepare zinc MOF.

Solid state and solution ^{1}HNMR study on the cobalt MOF provides energy barrier in the solid state for shuttling motion of the crown ether as 14.1 kcal mol^{-1} whereas in solution it is 7.7 kcal mol^{-1}. The slower rate of shuttling in the solid state is due to the presence of a surrounding framework causing steric and/or electrostatic hindrances to the motion.

Shuttling mechanical movements in a rotaxane are utilized for carrying out catalytic reactions. For example a chemical between A-B with another reactant C leading to a product C requires assembling of a dissociation of A-B bond and formation of A-C bond. The switchable catalyst can be designed based on [2]rotaxane where position of a macrocycle is controlled by placing over a particular catalytically active sites. In a rotaxane bifunctional catalyst, in other word a rotaxane catalyst with two active binding sites to facilitate a reaction by binding and activating a substrate one site may be kept exposed covering another site covered with the ring of a rotaxane. Thus, concealing one of the substrate binding sites, forces activation to take place at the unconcealed site. Once, a substrate is activated and product is formed at a particular site of rotaxane, other site may be revealed by movement of the ring to original site. Such ring movements help to activate the other site, to either bind a substrate hence causes another activation process. Shuttle-like motion of a macrocyclic ring on a rotaxane thus is utilized to generate catalytic activity or to passivize the catalytic action. If there are two binding sites on the dumb-bell then a control movement of the macorcyclic ring can either expose or conceal the catalytic site by adopting a position over the catalytic site or at the other site as illustrated in Figure 3.31. Consider a rotaxane having secondary ammonium site which is catalytically active and another quaternary ammonium site which is catalytically in active. Such effect for catalyst "on" state can be generated by treating rotaxane comprising of **3.50** and **3.51**. Under this situation the ring is located over the heterocyclic quaternary salt and catalyst amine site is available for catalysis. Whereas, the same rotaxane **3.52a** on treatment

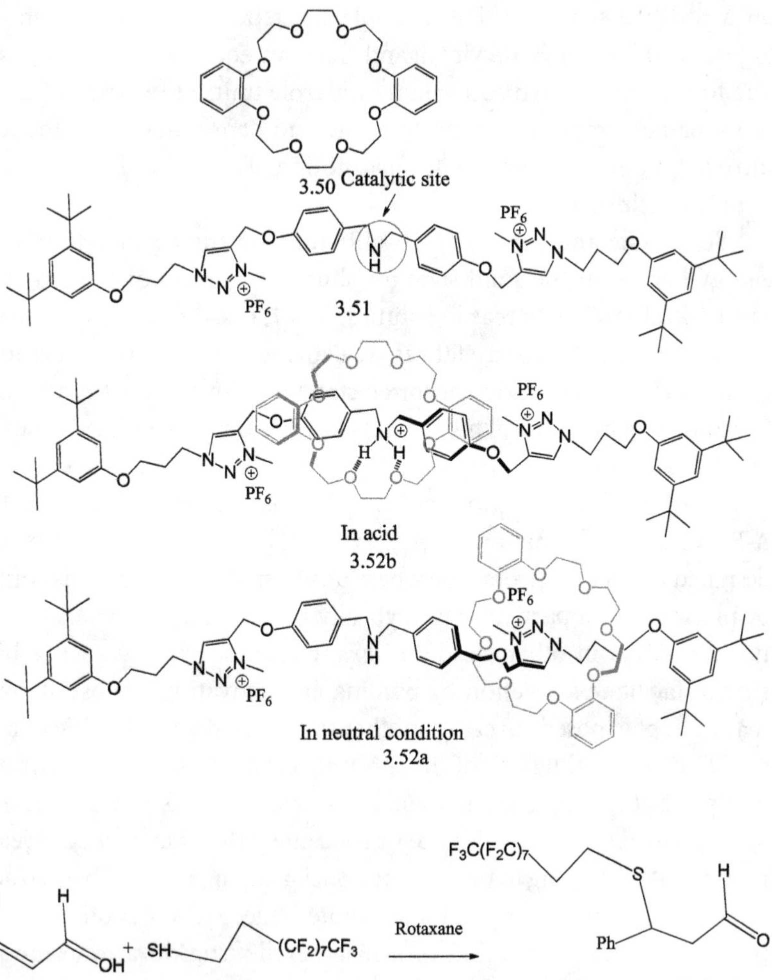

Figure 3.31.   Ring (**3.50**), rod (**3.51**) component of a rotaxane catalyst having dibenzylamine active site. **3.52a** and **3.52b** respectively are catalytically active and inactive states. Equation at the bottom of the figure is a representative reaction for such catalysis.

with acid a protonated state **3.52b** is generated. In acidic medium the macrocycle is held at this site to make a catalytically "off" state. Hence catalytic activity is lost in acidic condition. This catalyst is used in Michael addition of an aliphatic thiol to *trans*-cinnamaldehyde.

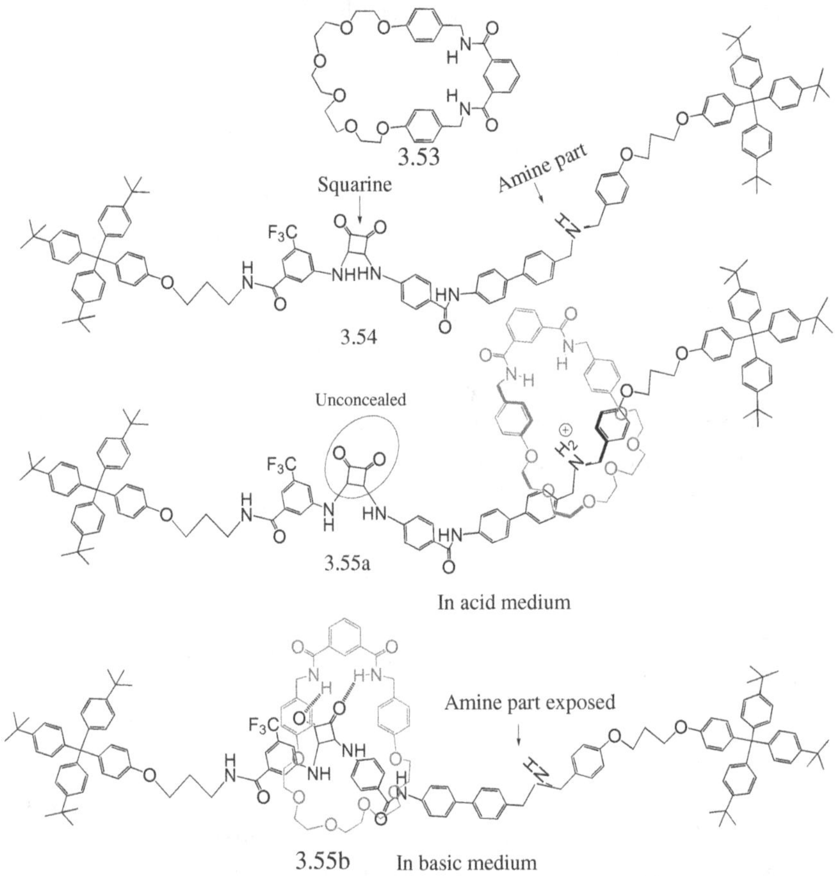

Figure 3.32. Ring (**3.53**), rod (**3.54**) of a catalytically active rotaxane with two active sites. States in which dibenzylamine site is concealed (**3.55a**) and squaramide part is concealed (**3.55b**).[20]

A rotaxane **3.55a** comprising of a rod that bears dibenzylamine and squaramide units at two well separated positions can be utilized for catalytic reactions such as Michael addition reaction. The rotaxane shown in Figure 3.32 has two binding sites with two bulky tri-4-tertiraybutylphenyl groups which promotes iminium catalysis. At other site having a squaramide part can activate electrophile by forming hydrogen bonds. On the other hand, the cyclic ring has a pyridyl-2,6-dicarboxyamide unit on one side and other side has flexible aliphatic ether linkages. The ring can form hydrogen bond selectively with squaramide by complementing hydrogen bond associated with pyridyl-2,6-dicarboxyamide unit. Besides these the polyether part of the crown ether-like region has a very high affinity for ammonium ions. Thus if the dibenzylamine unit is protonated it binds to the ammonium cation generated at this site, however neutral diben-zylamine will have very low affinity for the crown ether-like portion of the ring. Presence of a rigid spacer between the two active sites on the dumb-bell rod would avoid folding and facilitate shuttling movement of a ring. Thus these qualifications make the rotaxane suitable to perform shuttling motion back and forth in presence of acid followed by treatment with base. This rotaxane when treated with trifluoromethanesulphonic acid the amine part is protonated and the macrocycle preferentially encapsulate the dibenzylammonium group. This conceals the dibenzylamine for perform-ing catalysis. Thus any reaction catalyzed by iminium ion will not be possible under such a condition. Due to movement of the ring to conceal the dibenzylammonium site, the squaramide site becomes readily avail-able site. This allows squaramide part to be active site for performing catalytic action. Such catalysis occurs by activating a substrate by hydrogen bonds facilated by changing polarity of bonds of participating substrate. Under neutral condition the carbonyls of squaramide part are suitable hydrogen bond acceptors to bind with the macrocyclic ring, and the squaramide part is unavailable for catalysis and under this condition the amine part can act as catalytic site. This rotaxane promotes Michael addi-tion reactions through either part as active catalytic site. Any of these active sites can be made active or passive by controlling the acid-base in solution. Michael addition of 1,3-diphenylpropan-1,3-dione with croton-aldehyde or *trans-β*-nitrostyrene in presence of the rotaxane under different conditions helps to utilize the catalyst sites, with good selectivity. The machine action of this process can be described as ability to select the specific components of a mixture to react together, to form different prod-ucts from a common set of building blocks.

Circular and translational motions are generally required in complex machines. A model shown as **3.56** has two circular rotatable plates placed over balls. Each plate has arms of different kinds. One arm has a ring and other arm another circular plate has a rod with a ball attached at an end.

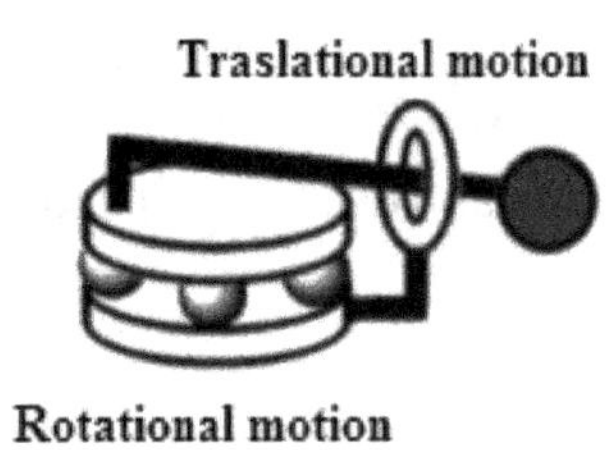

Traslational motion

Rotational motion

3.56

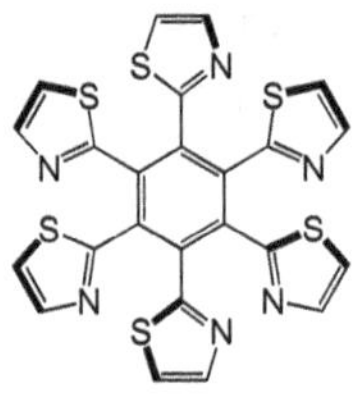

3.57

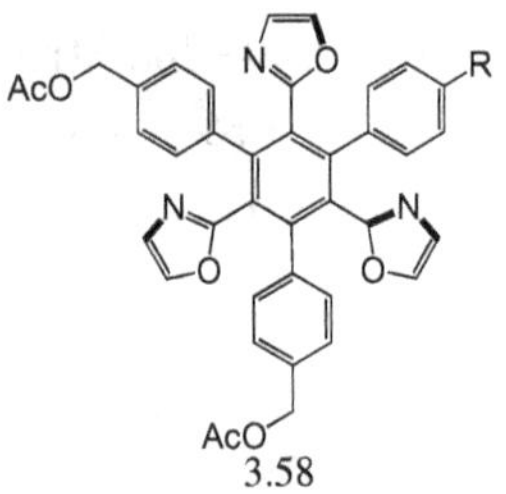

3.58

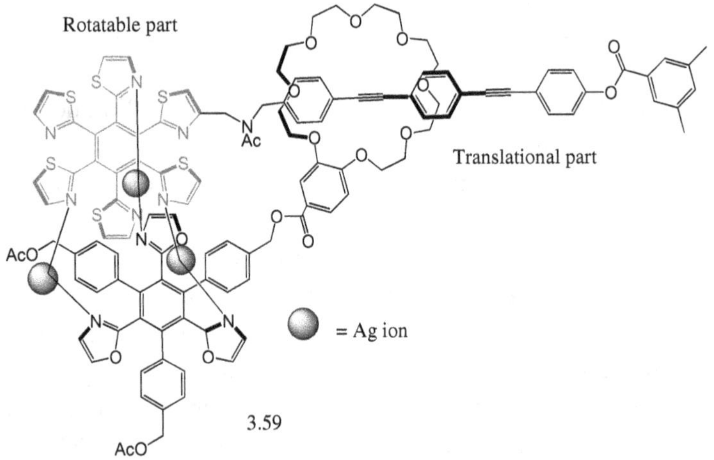

Rotatable part

Translational part

= Ag ion

3.59

Figure 3.33. Design of a molecular machine to execute translational and rotational motions.[21]

The arms ore arranged such that the rod is inserted into the ring. A circular motion of the plates over the balls causes the rod-like arm to move backward and forward giving a translational motion. Similar picture in molecular system comprising of silver complexes of derivatives of two compounds **3.57** and **3.58** are observed (Figure 3.33). These two molecules coordinate independently to silver ion and forms complex. Such silver complex replicates sandwiched metal ions which resembles ball sandwiched by two circular plates placed over each other. One ring attached to a long rigid part containing acetylenic unit may be considered as the rigid rod-like arm attached to a plate. A macrocycle attached to another plate-like structure can serve as the ring to insert the rod. Thus the molecule **3.59** is a model compound which show translational and rotation motion in a combined manner by complexation and decocomplexation with silver ions.

3.7 Knots

Knots are common topological features routinely applied in macroscopic world. A knot in mathematics means a topological state of a closed loop. In day to day life knots are used in wearing clothes, shoes, decoration and to connect between multiple sites or to keep united multiple components. Medical surgical sutures require concept of knot (Figure 3.34a) and knots are commonly come across to use in tying shoelaces (Figure 3.34b). The utility of knots is very common in every sphere of life. There are infinite numbers of ways to construct knots. A knot or knots may be present on a single thread, or between multiple threads or between closed tread with open single stretched thread or within a ring or in combination of rings. Two examples on construction of knots by a rope over a fixed rope are shown in Figure 3.35. Knots formed by vain can stop blood circulation; a knot formed by the umbilical cord during human pregnancy is dangerous. Knot-like arrangements of polymeric bio-molecules such as DNA, RNA and proteins are routine. To a chemist generation of knot-like architecture is a challenge. For a synthetic chemists design of molecular knot-like arrangements from molecules in predictable manner is a difficult task. On the other hand, knot-like structures provides avenues to understand the differences in properties of different states from a simple linear or unfolded form.

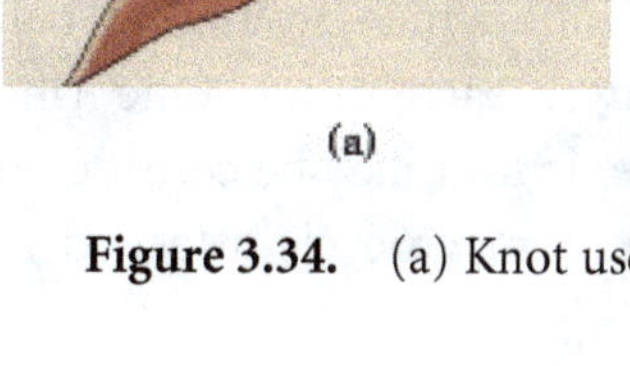

(a) 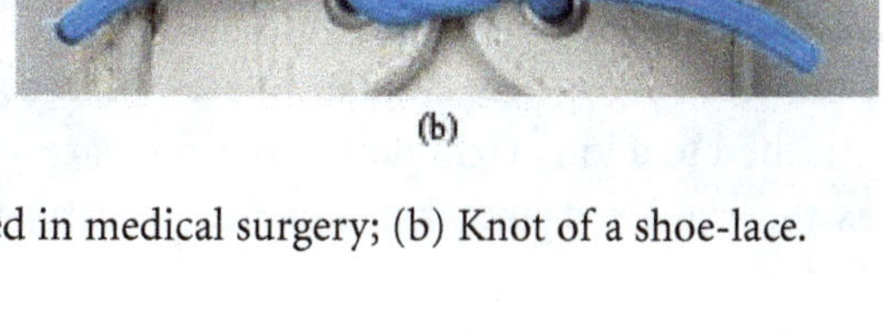　　　　　　　　(b)

Figure 3.34.　(a) Knot used in medical surgery; (b) Knot of a shoe-lace.

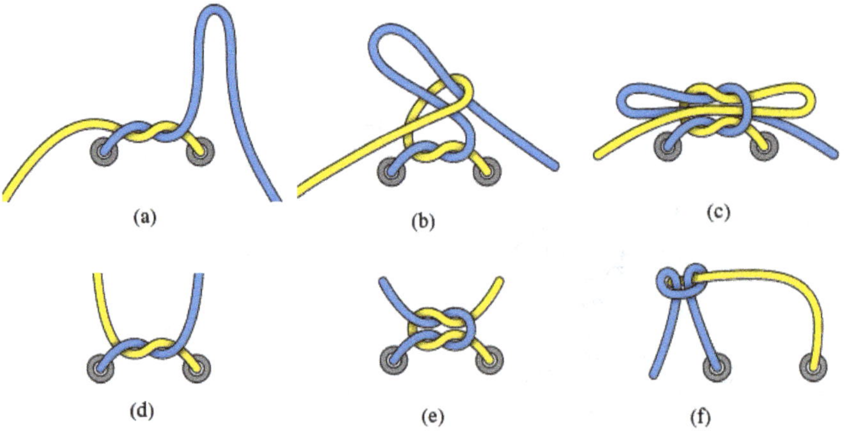

Figure 3.35. (a-c) Construction of a reversible knot; (d-e) construction of an irreversible knot, (f) Difficulty in recovering original state from an irreversible knot.

When two free ends or one end of a knot is pulled, it may result in a linear string like texture, in such a situation knot is called a reversible knot.[22] Same concept of reversibility can be extended to a loop producing a knot that is reverted to original closed curvature on pulling from one or multiple sites. Obvious reason for observing different types of knots is due to possibility of multiple ways for folding and tying while forming a knot. One terminology used to describe knot is called Alexander polynomial. In this classification knots are named according to the minimum number of crossings in a projection of the chain onto a plane. For such a classification a number followed by another subscripted number where both the numbers within a first bracket is used. First number is the crossing number and the subscripted number the order of the knot to differentiate it from amongst all knots with that crossing number. Notation (0_1) means no crossings which is referred to as unknot, a (3_1) knot has three crossings. Similarly, (4_1) knot has four crossings; accordingly there can be two knots with five crossings $(5_1, 5_2)$ and three knots with six crossings $(6_1, 6_2, 6_3)$. Mirror images of many knots are not superimposable, hence these knots are chiral, for example (3_1) knot is a chiral knot. A chiral knot can be converted to their mirror images. Knots are also classified as *torus* or *twist*

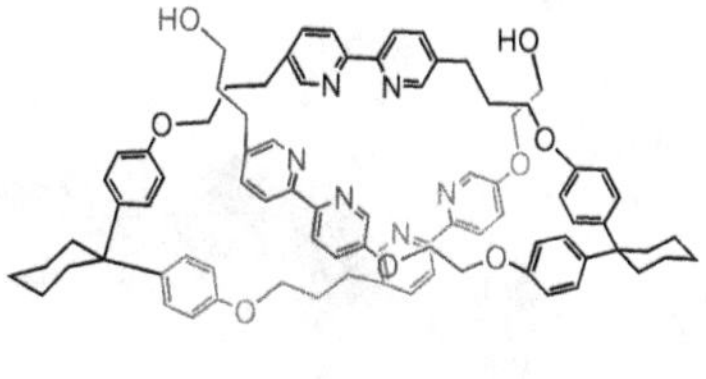

Figure 3.36. An organic compound **3.60** suitable to form knot **3.61–3.63** are some zinc complexes as knots.[23]

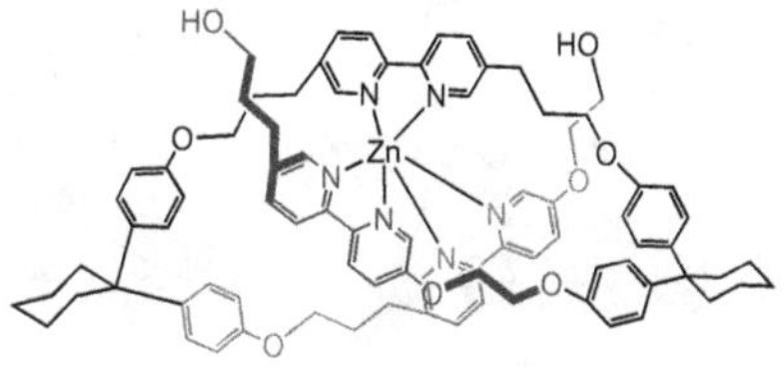

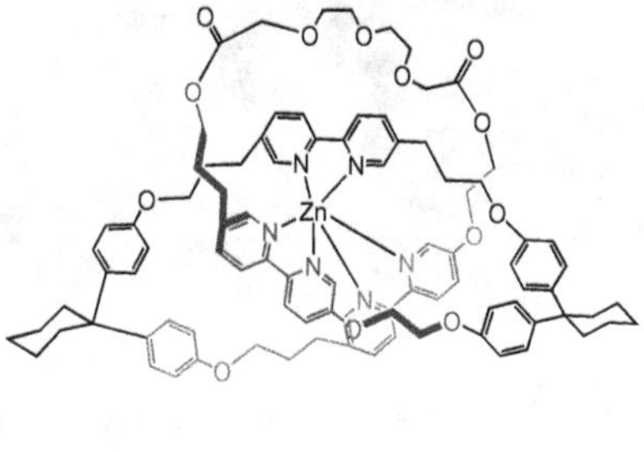

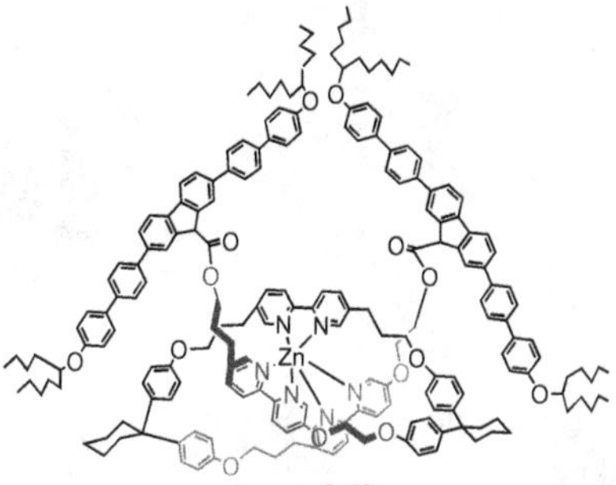

knots. Knots such as 3_1, 5_1, 7_1 can be drawn as closed curves on the surface of a torus, hence are called as Torus knots. Whereas twist knots are formed by linking the ends of a repeatedly twisted, closed loop and examples are 4_1, 5_2, 6_1 knot.

Organic supramolecular or inorganic complex are used as templates to organise flexible long chain compounds having multiple binding sites or function groups for synthesis of knots. In such templates end groups of folded molecules are connected by reacting among themselves or by another difunctional compound. In inorganic template, finally metal ions are removed to obtain desired knot. In this example the ligand **3.60** is suitable for formation of a knot. In the example shown in Figure 3.36 the compound **3.60** utilises the coordination effect of zinc ion to form knots. The ligand has three chelating sites to bind zinc ions, and it forms a six-coordinate complex which is in the form of a knot. This knot is easily dislodged by hydrolysis of ester groups. Another example **3.62** is example of a reversible knot formed as zinc complex, it goes back to stretched form of the ligand upon treatment with chloride ions.

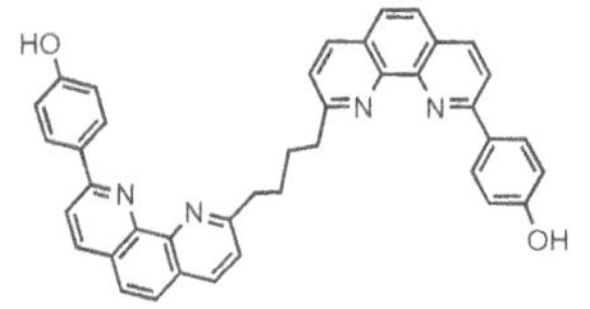

Figure 3.37.   A ligand used for knot and corresponding inorganic template mediated formation of knot.[24]

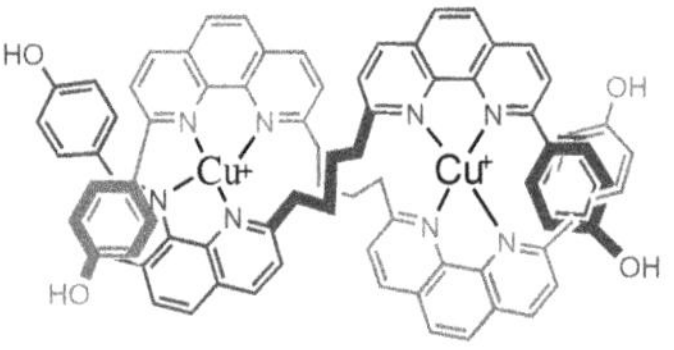

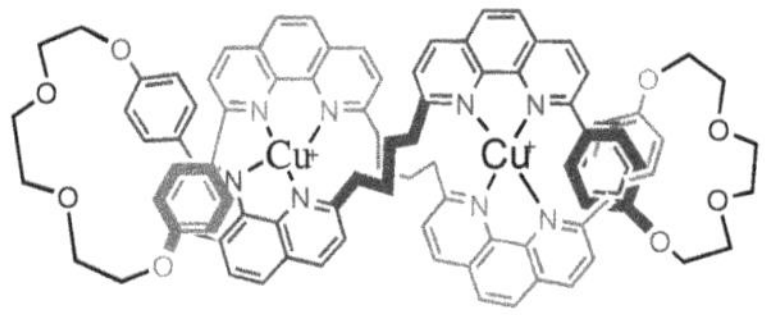

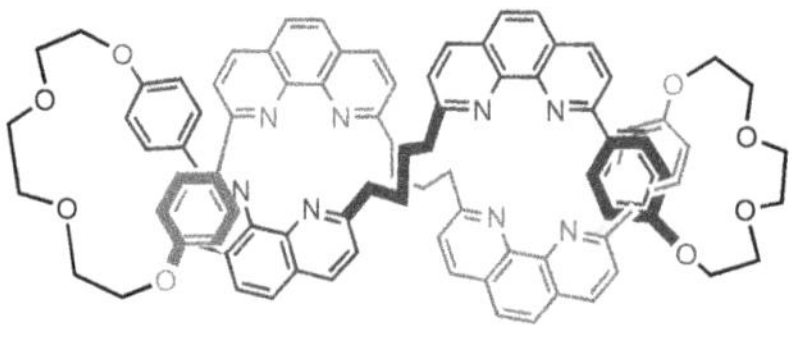

A template for a knot may be based on multiple binding sites to accommodate more than one metal ions (Figure 3.37). 1,10-Phenanthroline derivative **3.64** is used to prepare bimetallic copper(I) complex **3.65**. The complex **3.65** has a puckered template which is closed at the both phenolic ends by cross-linking to from crown ether type structure **3.66**. Removal of copper ions from the template provides a knot **3.67**. In such metal ion templating process, the ligands folds and form compact structure leading to contraction and upon removal of metal the knot opens up to adopt a more flexible structure, process is like contraction and expansion.

3.8 Controllable Properties in Rotaxanes Relating to Molecular Machines

Metal complexation may be used to stop equilibration of a ring in two equivalent sites of a rotaxane. Example illustrated in Figure 3.38 shows that a bipyridinium cation containing rotaxane **3.68** equilibrating between two equivalent positions to show another equivalent form. Equilibration is stopped by coordinating the 2,2′-bipyridine of the axle to copper ion. Copper complex **3.69** thus formed upon treatment with ion-exchange

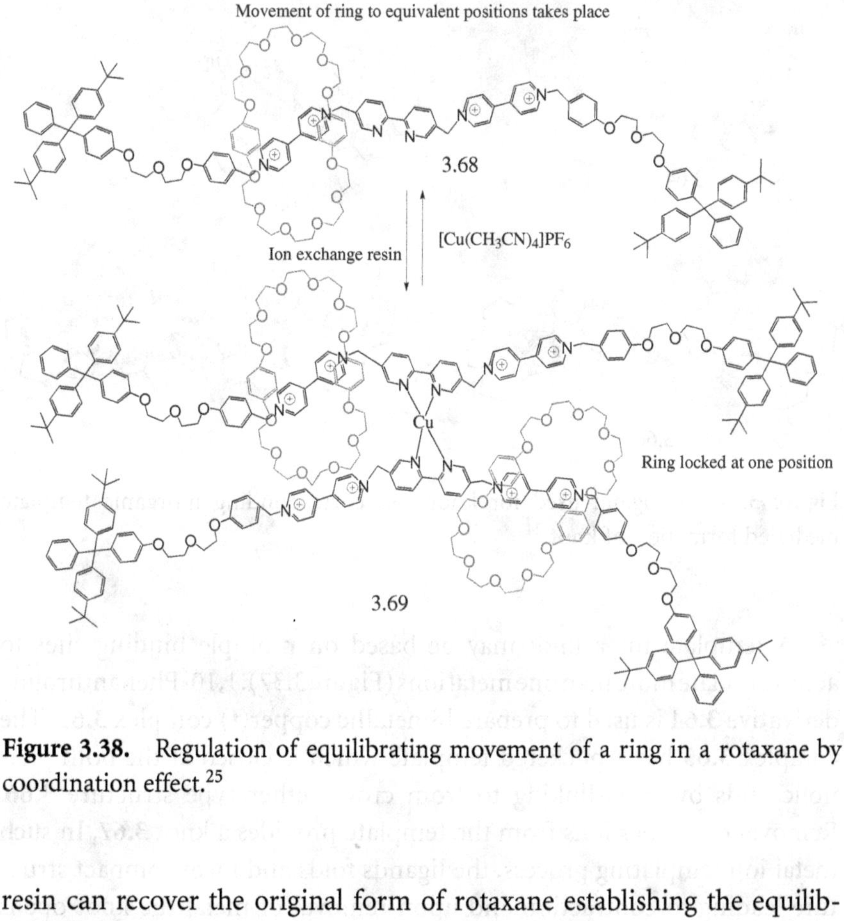

Figure 3.38. Regulation of equilibrating movement of a ring in a rotaxane by coordination effect.[25]

resin can recover the original form of rotaxane establishing the equilibrium again.

3.9 Molecular Machine for Peptide Synthesis

In biology proteins are formed by combination of amino acids in an orderly determined manner. Orderly sequences of peptide bonds from a set of specific amino acids are governed by messenger RNA. Rotaxane can be utilized to make systematic polypeptide formation by utilization of the shuttling motion as guiding factor. Operation of a ring acting as a machinery of an artificial small-molecule machine over a rod-like structure travelling along a molecular strand, can help in preparing very well

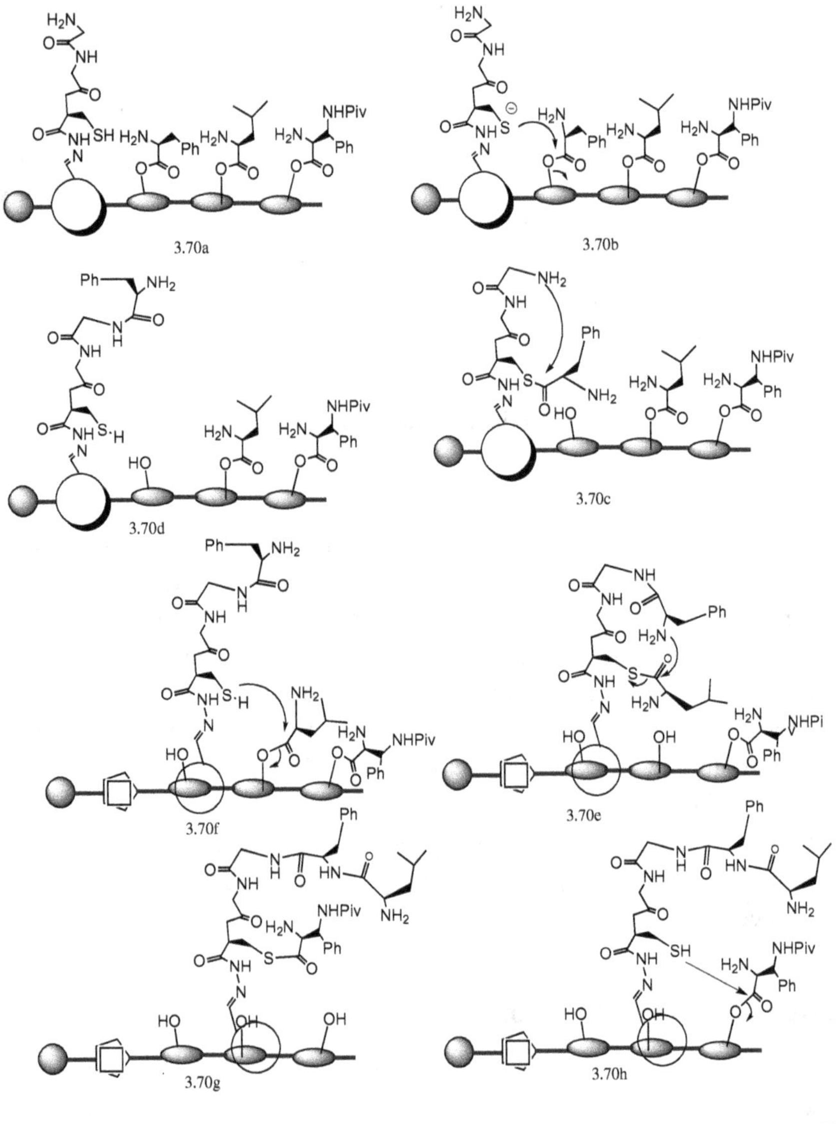

Figure 3.39. Path involved in rotaxane based machine to synthesise polypeptide.[26]

organized peptides. While movement of a small ring over the rod it picks up amino acids that block its path. This provides a means to synthesize a peptide in a sequence-specific manner (Figures 3.39 and 3.40).

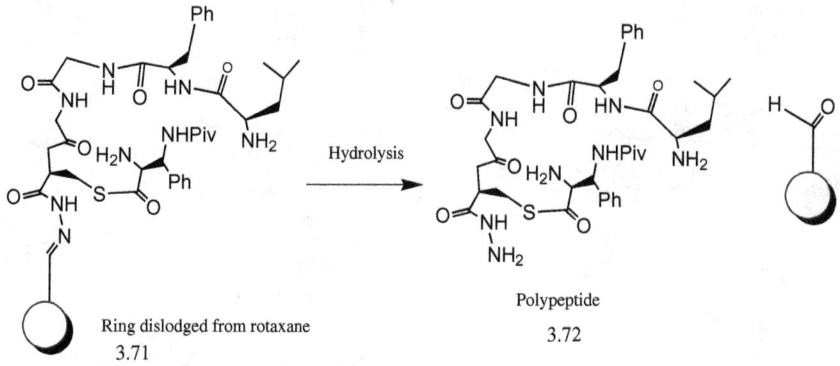

Figure 3.40.   Dislodging of raotaxane to get final product.

One such synthetic process involving a strand bearing amino acid building blocks and a macrocycle with a site for attachment of the reactive arm is shown in Figure 3.39. In this example a terminal blocking group is used to prevent the threaded macrocycle from coming off the strand. Reaction of thiol with caboxy-ester attached to macrocycle helps the transfer of amino acid sequentially placed over a strand. The amino group at a remote end reacts to regenerate the thiol end. This process repeats until all of the amino acid groups are cleaved (Figure 3.39). Finally the amino acid group containing polypeptide gets dislodged. Such dislodged compound undergoes hydrolysis to provide the desired sequentially designed polypeptide.

3.10 A Molecular Elevator

Combination of macrocycles to a core unit is used as a basis to understand the nature of inter-component electronic interactions. One such crown **3.73** undergoes assembling and dissembling with tripodal guest **3.74** in presence of acid or base (Figure 3.41). This 1 : 1 adduct 3.74, in which three crown rings of **3.73** holds the trifurcated guest **3.73** with the aid of three dibenzylammonium ions. The central core of the crown is fluorescent and the binding process can be evaluated by studying the fluorescence process. A solution of the adduct dissolved in acetonitrile undergoes dethreading-

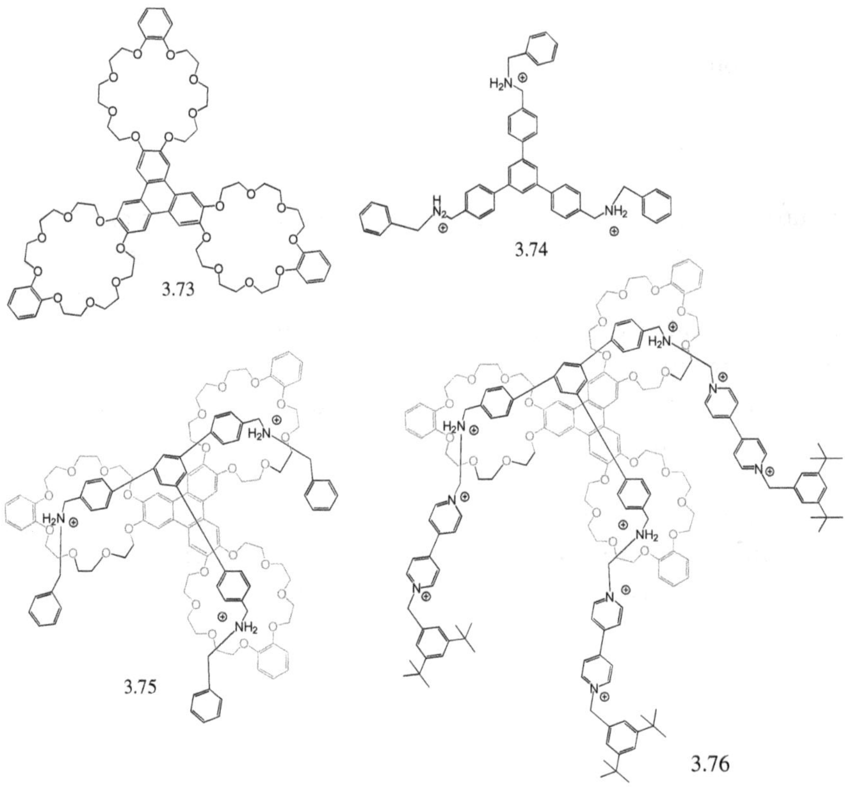

Figure 3.41. Principle for a molecular elevator.[27]

rethreading; which is readily controlled by adding base and acid. By sequential addition assembling and disassembling can be tuned. This principle is extended to make molecular elevators. In the case of an elevator a pyridinium cation is introduced on the host so that it provides two sites on each arm of the tripodal guest. The host-guest complex has now possibility to hold crowns at two different sites of each arm. This nano-machine, is of about 2.5 nm height with a diameter about 3.5 nm. It can be tuned at two positions by causing upward or downward movement. Adjustment of pH to neutral condition to convert benzylammonium cation to benzyl amine pushes the ring to take position over bipyridinium cationic units. On the other hand, upon protonation of benzylamine the ring shifts to the site where benzylammonium cations are located. The "up and down" elevator-like motion, can be performed repeatedly for several times.

3.11 Logic Gate Based on Rotaxane

Rotaxane **3.77** shows properties of logic gates by acting as an array of configurable electronic switches (Figure 3.42). The switches are generated by forming monolayer of redox-active rotaxane units sandwiched between metal electrodes. The property of logic gate is established by monitoring current flow at reducing voltages. To show such properties electronically configurable junction having molecular monolayer sandwiched between lithographically fabricated metal wires are used. Such junction act as switch, and combination of several such arrangements, provides scope to perform as electronically configurable wired-logic gates. In addition to several advantages over conventional devices used for similar purposes, these devices have other advantage that they can be miniaturized to molecular dimensions.

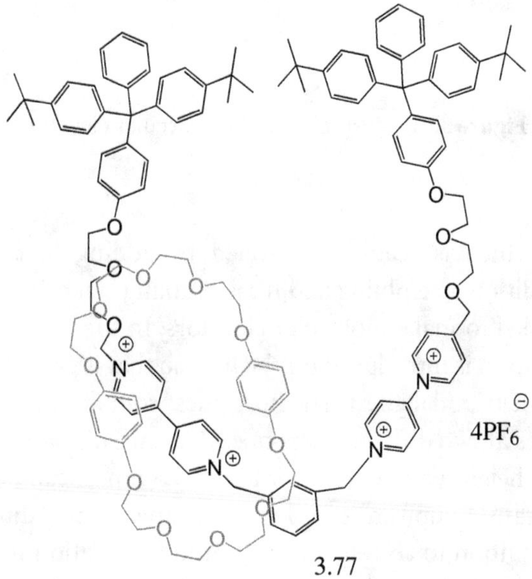

3.77

Figure 3.42.   A rotaxane used for making electronic logic gate.[28]

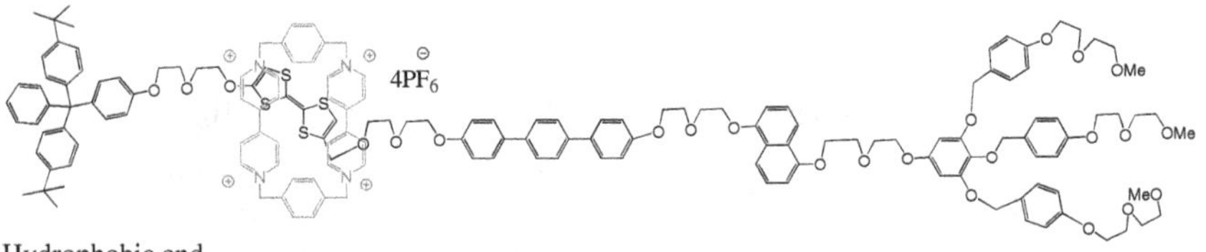

Hydrophobic end

Hydrophilic end

4PF$_6$

3.78

Figure 3.43. A rotaxane used for molecular electronic memory device.[29]

The rotaxane **3.78** has bulky hydrophilic and hydrophobic stoppers at two ends and has three distinguishable stations to binding to a tetracationic ring unit comprising of 4,4′-bipyridinium cations. The ground state of the rotaxane has ring unit embedding tetra-thiofulvalene unit as illustrated in Figure 3.43. When the hydrophilic stopper of the rotaxane is in contact with a Si bottom-nanowire electrode this state of the rotaxane shows low-conductance. Upon oxidation tetrathiafulvalene site generate mono or dication of tetrathiofulvalene ion. Generation of such a cation causes translation motion of the ring to the dioxynaphthalene site. Cationic tetrathiofulvalene on reduction to neutral state form a metastable co-conformer which shows high-conductance. Such a metastable state slowly returns to the ground state which is insulating state. Based on such observations a 160–kilobit molecular memory device was patterned.[30] There are serious efforts on new product design[31] and to use them in operation of objects of macroscopic world.[32–33]

References

1. Laidler DA, Stoddart JF (1977) Stereoselectivity in complexation of primary alkylammonium cations by the diastereotopic faces of chiral asymmetric crowns, *J. Chem. Soc. Chem. Commun.* **14**: 481–483.
2. Curtis WD, Laidler DA, Stoddart JF, Wolstenholme JB, Jones GH (1977) Enantiomeric differentiation by a chiral symmetrical crown derived from L-iditol, *Carbohydrate Res.* **57**: C17–C22.
3. Collin J-P, Heitz V, Sauvage J-P (2005) Transition-metal-complexed catenanes and rotaxanes in motion: Towards molecular machines. *Topics in Current Chem.* **262**: 29–625.

4. Chambron JC, Harriman A, Heitz V, Sauvage JP (1993) Ultrafast photoinduced electron transfer between porphyrinic subunits within a bis(porphyrin)-stoppered rotaxane. *J. Am. Chem. Soc.* **115**: 6109–6114.

5. Anelli PL, Spencer N, Stoddart JF (1991) A Molecular Shuttle. *J. Am. Chem. Soc.* **113**: 5131–5133.

6. Bissell RA, Cordova E, Kaifer AE, Stoddart JF (1994) A chemically and electrochemically switchable molecular shuttle. Nature **369**: 133–137.

7. Anelli PL, Spencer N, Stoddart JF (1991) A molecular shuttle. *J. Am. Chem. Soc.* **113**: 5131–5133.

8. Silvi S, Arduini A, Pochini A, Secchi A, Tomasulo M, Raymo FM, Baroncini M, Credi A (2007) A simple molecular machine operated by photoinduced proton transfer. *J. Am. Chem. Soc.* **129**: 13378–13379.

9. van Delden RA, ter Wiel MKJ, Koumura N, Feringa BL (2003) Synthetic molecular motors. *Topics in Current Chem.* **262**: 559–577.

10. Chatterjee MN, Kay ER, Leigh DA (2006) Beyond switches: Ratcheting a particle energetically uphill with a compartmentalized molecular Machine. *J. Am. Chem. Soc.* **128**: 4058–4073.

11. Iijima T, Vignon SA, Tseng H-R, Jarrosson T, Sanders JKM, Marchioni F, Venturi M, Apostoli E, Balzani V, Stoddart JF (2004) Controllable Donor–Acceptor Neutral [2]Rotaxanes. *Chem. Eur. J.* **10**, 6375–6392.

12. Asakawa M, Brancato G, Fanti M, Leigh DA, Shimizu T, Slawin AMZ. Wong JKY, Zerbetto F, Zhang SW (2002) Switching "On" and "Off" the expression of chirality in peptide rotaxanes. *J. Am. Chem. Soc.* **124**: 2939–2950.

13. Mateo-Alonso A, Fioravanti G, Marcaccio M, Paolucci F, Jagesar DC, Brouwer AM, Prato M (2006) Reverse shuttling in a fullerene-stoppered rotaxane. *Org. Lett.* **8**: 5173–5176.

14. Leigh DA, Morales MA, Perez EM, Wong JK, Saiz CG, Slawin AM, Carmichael AJ, Haddleton DM, Brouwer AM, Buma WJ, Wurpel GW, Leon S, Zerbetto F (2005) Patterning through controlled submolecular motion: rotaxane-based switches and logic gates that function in solution and polymer films. *Angew. Chem. Int. Ed.* **44**: 3062–3067.

15. Umehara T, Kawai H, Fujiwara K, Suzuki T (2008) Entropy- and hydrolytic-driven positional Switching of macrocycle between imine and hydrogen-bonding stations in rotaxane-based molecular shuttles. *J. Am. Chem. Soc.* **130**: 13981–13988.

16. Qu D-H, Feringa BL (2010) Controlling molecular rotary motion with a self-complexing lock. *Angew. Chem., Int. Ed.* **49**: 1107–1110.

17. Ashton PR, Ballardini R, Balzani V, Boyd SE, Credi A, Gandolfi MT, Gomez-Lopez M, Iqbal S, Philp D, Preece JA (1997) Simple mechanical molecular

and supramolecular machines: photochemical and electrochemical control of switching processes. *Chem. Eur. J.* **3**: 152–170.

18. Zhu K, O'Keefe CA, Vukotic VN, Schurko RW, Loeb SJ (2015) A molecular shuttle that operates inside a metal-organic framework. *Nature Chem.* **7**: 514–519.

19. Blanco V, Carlone A, Hanni KD, Leigh DA, Lewandowski B (2012) A rotaxane-based switchable organocatalyst. *Angew. Chem. Int. Ed.* **51**: 5166–5169.

20. Beswick J, Blanco V, Bo GD, Leigh DA, Lewandowska U, Lewandowski B, Mishiro K (2015) Selecting reactions and reactants using a switchable rotaxane organocatalyst with two different active sites. *Chem. Sci.* **6**: 140–143.

21. Okuno E, Hiraoka S, Shionoya M (2010) A synthetic approach to a molecular crank mechanism: Toward intramolecular motion transformation between rotation and translation. *Dalton Trans.* **39**: 4107–4116.

22. Danon JJ, Krüger A, Leigh DA, Jean-François Lemonnier JF, Stephens AJ, Vitorica-Yrezabal IJ, Woltering SL (2017) Braiding a molecular knot with eight crossings. *Science* **355**: 159–162.

23. D'Souza DM, Leigh DA, Papmeyer M, Woltering SL (2012) A scalable synthesis of 5,5′-dibromo-2,2′-bipyridine and its stepwise functionalization via Stille couplings. *Nature Protocols* **7**: 2022–2028.

24. Cárdenas D, Livoreil A, Sauvage J-P (1996) Redox control of the ring-gliding motion in a Cu-complexed catenane: A process involving three distinct geometries. *J. Am. Chem. Soc.* **118**: 11980–11981.

25. Jiang L, Okano J, Orita A, Otera J (2004) Intermittent molecular shuttle as a binary switch. *Angew. Chem Int. Ed.* **43**: 2121–2124.

26. Lewandowski B, Bo GD, Ward JW, Papmeyer M, Kuschel S, Aldegunde MJ, Gramlich PME, Heckmann D, Goldup SM, D'Souza DM, Fernandes AE, Leigh DA (2013) Sequence-specific peptide synthesis by an artificial small-molecule machine. *Science* **339**: 189–193.

27. Badjic JD, Balzani V, Credi A, Silvi S, Stoddart J F, (2004) Molecular elevator. *Science* **303**: 1845–1849.

28. Collier CP, Wong EW, Belohradsky M, Raymo MFM, Stoddart JF, Kuekes PJ, Williams RS, Heath JR (1999) Electronically Configurable molecular-based logic gates. *Science* **285**: 391–394.

29. Kay ER, Leigh DA, Zerbetto F (2007) Synthetic molecular motors and mechanical machines. *Angew. Chem. Int. Ed.* **46**: 72–191.

30. Green JE, Choi JW, Boukai A, Bunimovich Y, Johnston-Halperin E, DeIonno E, Luo Y, Sheriff BA, Xu K, Shin YS, Tseng H-R, Stoddart JF, Heath JR (2006) A160–kilobit molecular memory electronic memory patterned at 10^{11} bits per square centimeter. *Nature* **445**: 414–417.

31. Collier CP, Wong EW, Belohradsky M, Raymo FM, Stoddart JF, Kuekes PJ, Williams RS, Heat JR (1999) Electronically configurable molecular-based Logic gates. *Science* **285**: 391–394.

32. Berna J, Leigh DA, Lubomska MS, Mendoza M, Perez EM, Rudolf P, Teobaldi G, Zebetto F (2005) Macroscopic transport by synthetic molecular machines. *Nature Mater.* **4**: 704–710.

33. Eelkema R, Pollard MM, Vicario J, Katsonis N, Ramon BS, Bastiansen CW, Broer DJ, Feringa BL (2006) Molecular machines: nanomotor rotates microscale objects. *Nature* **440**: 163.

4

Photochemically and Electrochemically Guided Molecular Machines

4.1 Principles of Photochemical Switching

Artificial motors run on chemical as fuel cannot be considered as autonomous. This is because of the fact that after every mechanical motion caused by an external chemical reagent there is need of another reagent to cause or retrace the mechanical movement. Thus in such process there is accumulation of waste products and additional process to destruct the waste product is essential. But there are other forms of energy that do not necessarily generate waste product. One of the energy sources routinely used by nature is the light energy. Light is comprised of photons and can be easily obtained in a particular wavelength which is called monochromatic light. Photochemical energy can be a source to stimulate mechanical movement in compounds. Such changes are common in *cis-trans* photo-isomerization reactions involving –N=N–, –C=N– or –C=C– double bonds. Advantage of the geometrical changes across a double bond is very attractive as such reactions maintain atom economy and are reversible reactions. The inter-conversion between *cis and trans* isomers causes light-driven operation of molecular machines.[1-3] It is possible due to structural changes causing motions with large amplitude depending on the substrate under consideration. The isomeristion can be brought about in molecules having crown ethers attached at two sides of an azo-group (Figure 4.1). In such case irradiation with suitable wavelength causes molecular motions.

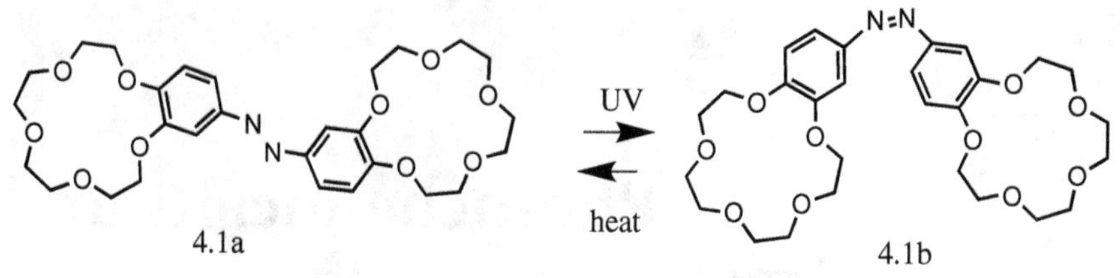

Figure 4.1. *Cis and trans* form of a *bis*-crown ether linked by azo group.[4]

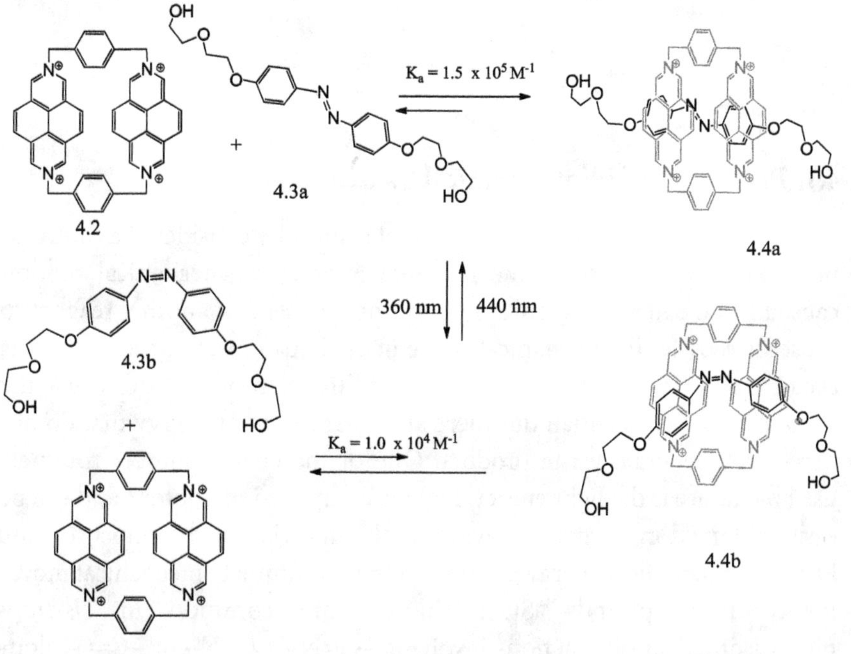

Figure 4.2. *Cis-trans* equilibrium caused by photo irradiation.[5]

Another example of such a rotaxane having an azo group on axle component is illustrated in Figure 4.2. Trans form of molecule **4.3a** can form rotaxane with a ring **4.2** with equilibrium constant $1.5 \times 10^5 \, M^{-1}$, whereas *cis* form **4.3b** form with a lower equilibrium constant $1.0 \times 10^4 \, M^{-1}$. Thus, on irradiation at 360 nm axle of the rotaxane adopts a *cis* geometry but the concentration of rotaxane in *cis* form is reduced due to the lower

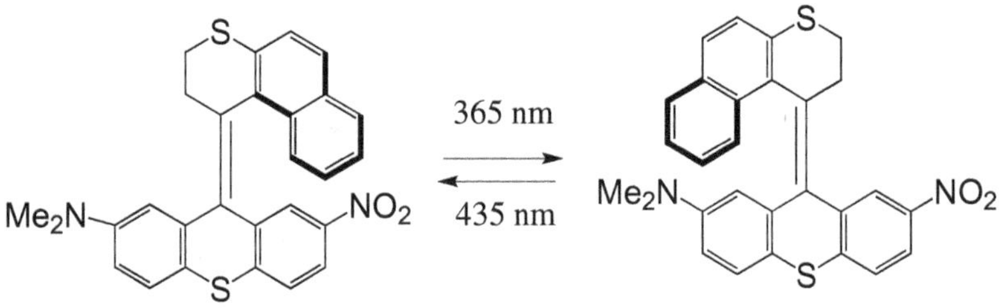

365 nm
⟶
←
435 nm

Me₂N —   NO₂

4.5a   4.5b

Figure 4.3.   Light-driven changes in orientations across a double bond.[6]

equilibrium constant and the shift of equilibrium towards free states. On the other hand the *cis* form can be converted to *trans* form by irradiating at 440 nm and the concentration of *trans* form increases in solution over the rotaxane in *cis* form. Equilibration between different geometrical structures due to conversion of *cis*-azo group to *trans*-azo group influences binding affinity of crown ether.

Photo-isomerisation by choosing adequate wavelengths causes reversible change across double bond in unsymmetric chromophoric compounds to act as a switch (Figure 4.3). The compound **4.5a** has a three fused-ring system as stator which is connected to another three fused ring system though a double bond acting as a rotor. Orientations of naphthalene part can be altered across the stator by irradiation with appropriate wavelength.

Photoisomerizations about a –C=C– bond of sterically hindered alkenes can be utilised in construction of molecular rotary motors. For example *cis-trans* conversions are possible in different ways across a double bond to generate different regiomers as illustrated in Figure 4.4. In this compound due to axial and equatorial positions of the methyl groups on the cyclohexane ring four states can be generated. Among them the one having *cis* regio-isomer with methyl groups in equatorial and equatorial positions is less stable than the *cis* form of regiomer with methyl groups at *trans* and *trans* positions. Hence former form easily changes to later isomer. On the other hand in the case of *trans* isomers also there are two states, one has both the methyl groups at equatorial positions which transforms on heating to another *trans* form that has both the methyl groups at

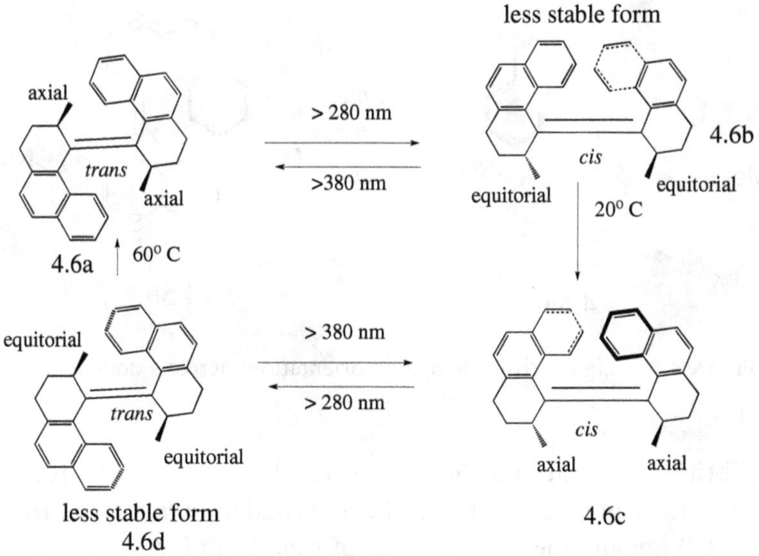

Figure 4.4. Different orientations of naphthalene acrosss a double bond by irradiations.[7]

trans positions. In this example four states of slightly different energy are equilibrated by using light as well as heat resulting in mechanical motion across a double bond of different forms 4.6a–d.

4.1.1 *Movement of cyclodextrin by photoisomerisation of rotaxane*

Cyclodextrin is a cyclic oligosacharaide and it is classfied as α, β, Υ depending on number of sugar units present as the ring of such cyclic oligomers. These compounds are made of hollow rim, one side of which is hydrophilic and other side is hydrophobic. Hence, advantage of passing hollow structure is that hydrophilic and hydrophobic effect can be utilised to cause movement of such cyclodextrin ring over molecules acting as axle.

Movement of cyclodextrin ring in a rotaxane can be achieved by causing *cis-trans* isomerisation in a rotaxane shown in Figure 4.5. Conversion of *cis-trans* geometry faciliates change of positions and in *trans* form cyclodextrin ring moves between thermodynamically stable positions.

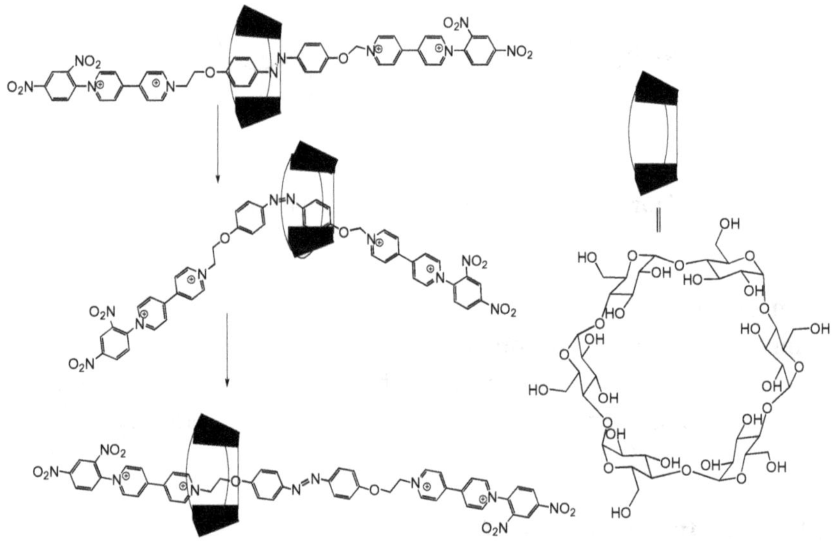

Figure 4.5.   Movement of a cyclodextrin part of a rotaxane upon irradiation.[8]

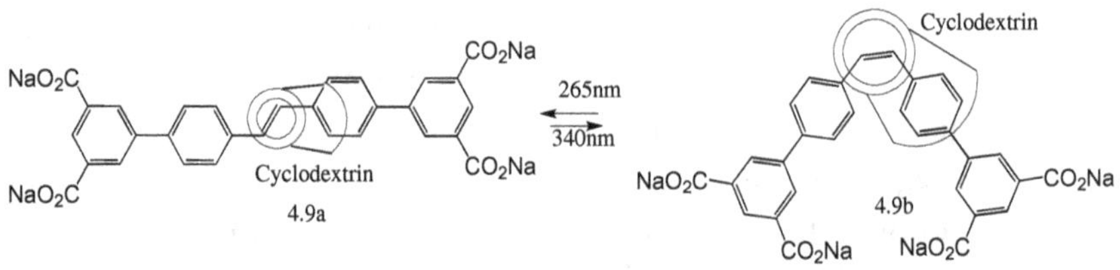

Figure 4.6.   Change in orientation of C=C of a rotaxane upon irradiation.[9]

In analogous manner *cis-trans* geometrical changes across a water soluble rotaxane (Figure 4.6) can be used to cause reversible motions. In this example sodium salts of the dicarboxylic acids help in enhancing solubility of the axle part in aqueous solvent.

4.1.2 *Photochemically guided keto-enol forms*

Light-driven transformations of enol form to keto form are utilized in reversibly interchanging enol and keto forms. Thiazole detivative connected

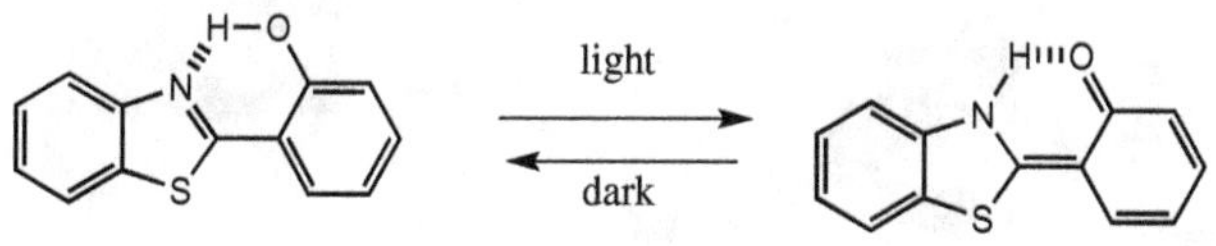

Figure 4.7. Keto-enol forms formed by irradiation.

to an ortho-hydroxy phenol (Figure 4.7) provides an example for such changes. The keto form of compound **4.10a** converts to enol form by light. Formation of intramolecular hydrogen bonds helps to stabilize keto form. Keto form is less thermodynamically stable than the enol form.

4.1.3 Utility of photoreduction in switching

Photoreduction or electron transfer processes are also successfully utilized to cause molecular motions. For example, charge-transfer interaction is possible between electron donor and electron acceptor units. Such interactions result in charge-transfer excited states which have comparatively low energy gap with respect to the corresponding ground state. Presence of such interactions provides relatively weak and broad absorption bands in the visible region. Presence of a charge-transfer interaction also causes quenching of other luminescent excited states which are localized on each component. On the other hand, partner molecules of a charge-transfer complex are more difficult to oxidize or reduce than individual partner molecules. Moreover, charge-transfer interactions guide mechanical movements of interlocked molecules such as in rotaxanes and catenanes. In other word, if one of the component or both components involved in charge-transfer interactions are oxidized or reduced the movement may altogether change than a species that originally had a charge-transfer interactions. Such redox processes may be induced chemically, electrochemically, or by photochemical means. Alternately, charge-transfer interactions can be introduced by redox reactions generating suitable oxidation states of partner molecules to have adequate charge-transfer. Such a process may be of utility in reversing a mechanical movement by combining redox reactions with charge-transfer interactions. Compound **4.11**

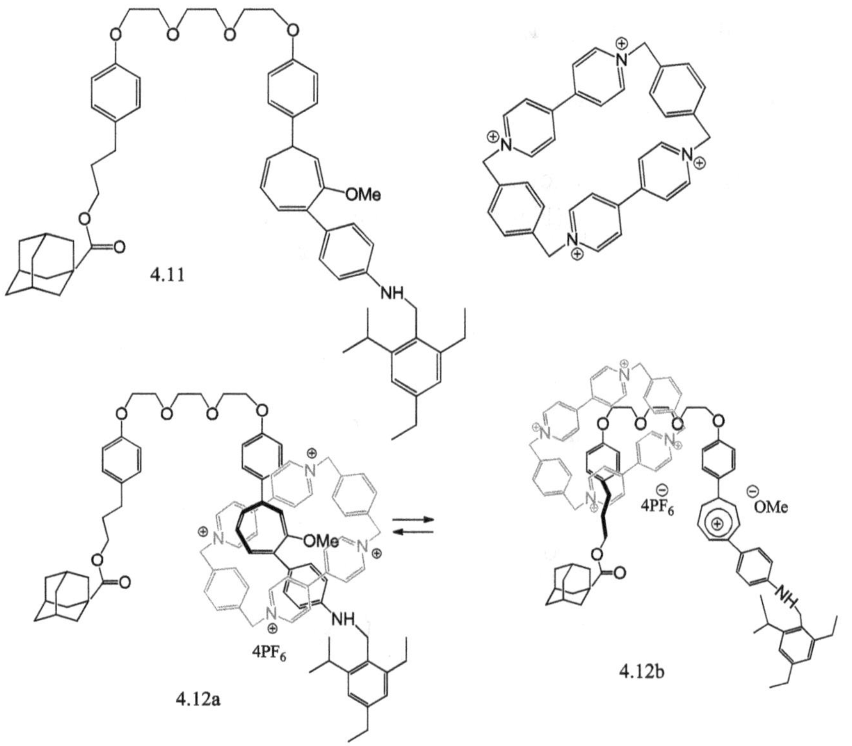

Figure 4.8. Movement of ring due to loss of charge-transfer interactions by generation of a cation.[10]

with a ring derived from 4,4 -bipyridinium dications forms rotaxane in which the ring is located over the hepta-nuclear ring of the axle due to charge-transfer interaction, the compound shows absorption peak at 380 nm. On exposure to light the methoxide ion is released and aromatic cation is formed (Figure 4.8). The compound thus loses change-transfer interaction and faces repulsion due to a similar charged cation generated in the seven member ring as that of the charges present on the bipyridinium cations that are part of the interlocking ring. This forces the ring to move to an alternate position. The new state has visible absorption at 580 nm.

Rotaxane (**4.15a**) electron donor ring (**4.14**) embedding the dumbbell component comprising of four operating sites namely (a) ruthenium(II) polypyridine complex, (b) two electron acceptor sites, (c) rigid spacer for movement of the ring, (d) a tetraaryl methane group at one end as a stopper.

The ruthenium complex part acts as site to capture light energy to trigger the system and also act as stopper of the dumb-bell. The stable translational isomer of rotaxane has the crown ether ring positioned at electron acceptor site 1 (Figure 4.9) as the presence of methyl groups on

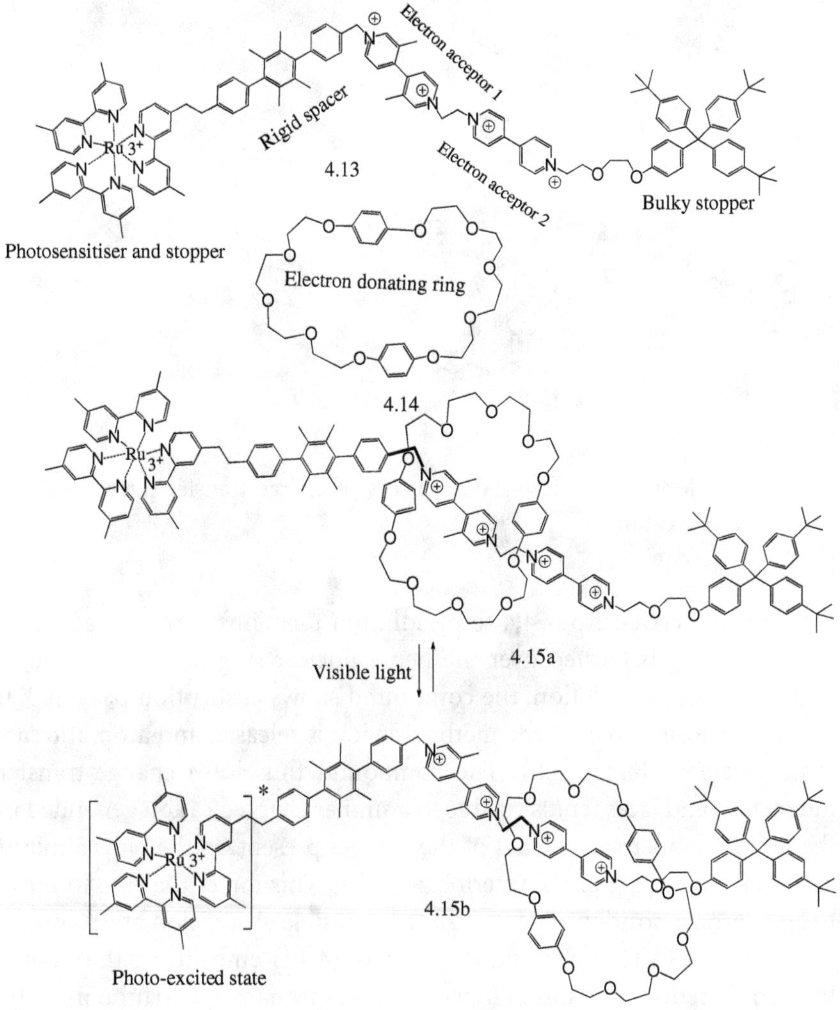

Figure 4.9. Photoinduced ruthenium complex assisted shuttling motion in a rotaxane.[11]

the bipyridium cations makes this relatively better electron donating site to interact with the crown, which is donor. This rotaxane shows photoin-duced abacus-like movement of the macrocycle between the two electron accepting sites upon irradiation. The movement of macrocyclic ring takes place due to excitation caused by light at the ruthenium center is trans-ferred to excited state of the acceptor 1. This causes deactivation of the site relative to the acceptor site 2. As a result of such a deactivation the donor crown ether ring moves to the position of acceptor 2. During the course of time, a back electron-transfer from the acceptor site 1 to the oxidized ruthenium complex takes place which restores the electron acceptor power of acceptor 1. This enforces a backward movement of the ring by Brownian motion. Solvent has a role in such movements, for example rate of ring shuttling in acetonitrile at room temperature is slower than the back electron transfer. Irradiation by visible light causing excitation at the ruthenium center cause a forward and backward ring movement at about 2 % quantum efficiency.

4.1.4 *Role of external photosensitisers*

For construction of information ratchet, there must be controlled way to open door approaching from a certain direction. Such positional sorting helps to accumulate a particular species (in the form of quanta of light or particle) in one side of the container. The rotaxane shown in Figure 4.10 is comprised of dibenzo-24-crown-8-based macrocycle. The crown ether unit is is mechanically locked on an axle by bulky 3,5-di-trertiarybutyl-phenylstoppers. The axle has α-methyl stilbene unit dividing the dumb-bell structure into two compartments. Both the compartments have ammo-nium binding sites. The crown ether has benzophenone unit which can be photosensitized. In this example photosensitized energy transfer from the macrocycle to the stilbene unit is used by the rotaxane to control the position of the macrocycle ring. E-isomer of stilbene helps to move the macrocycle randomly by Brownian motion within the ring; but Z iso-mer traps the ring at one of the compartments. Benzil (PhCOCOPh) is used as a sensetizer. It helps in formation of Z-isomer α-methyl stilbene. Irradiation in presence of benzil causes photoisomerisation results in Z:E isomers in 82:18 ratio. This executes a selective "gate opening" pro-cess by photosensitized energy transfer from the macrocycle to the

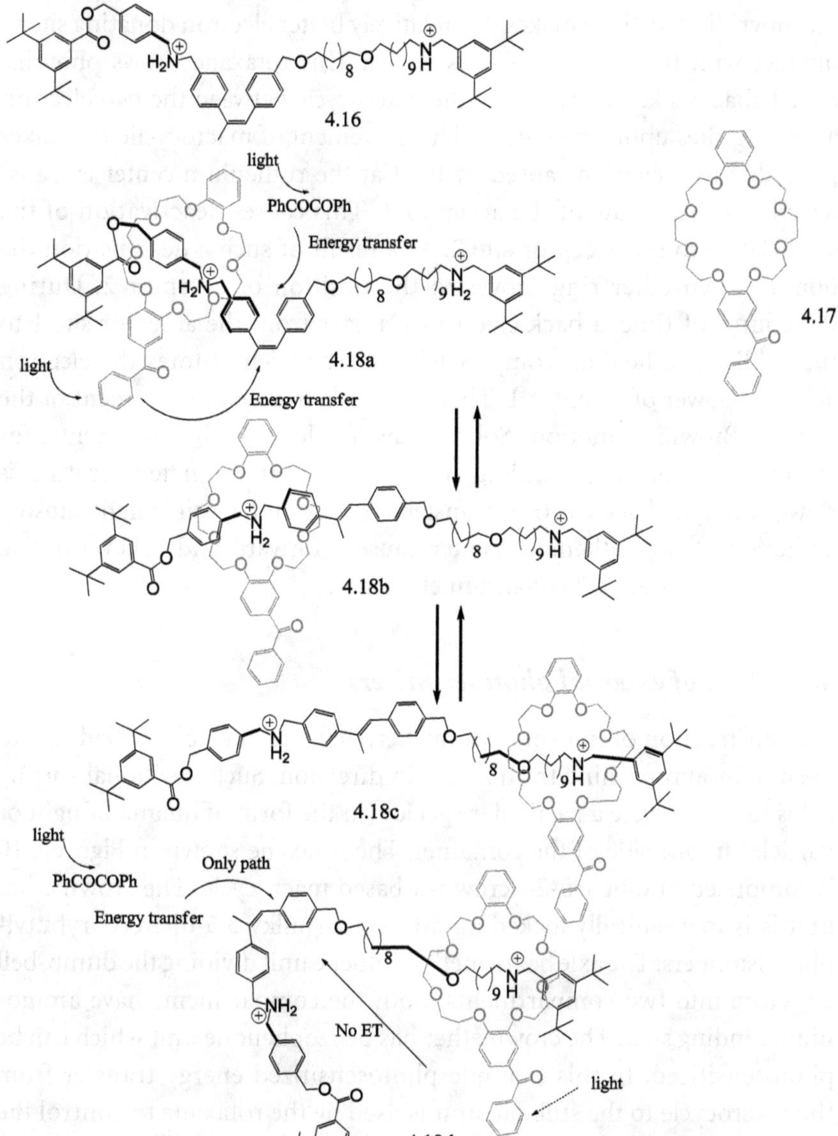

Figure 4.10. Photosensitization of a rotaxane by external or internal photosensitizer to change positions of rings.[12a]

methyl-stilbene unit. This led to 55:45 Z:E ratio of the α-methyl stilbene. Upon changes in isomer from *trans* to *cis* (E to Z) energy transfer is difficult due to distance between the sites from the benzophenone site. The reduction in intramolecular energy transfer from the macrocycle keeps the gate closed, which causes a bias on the distribution of macrocyclic ring over the axle.

4.1.5 *Photochemically driven molecular shuttle*

The example shown in Figure 4.11 is combination of three interlocked macrocyclic rings. The larger ring has several sites to interact with the smaller macrocyclic rings. Directional rotation with halts at different sites is achieved in this [3]catenane. Presence of two smaller rings helps each other to restrict the rotational freedom of the molecule. Large macrocycle has four different interaction sites among which two are fumaramide units having different binding abilities, one succinic amide ester unit, and an amide group. There is a benzophenone moiety attached at one end; which serves as sensitizer for the isomerization of the double bond of the fumaramide unit. The benzophenone unit helps photo-isomerization of olefinic unit through photo-energy transfer using a higher wavelength than a regular wavelength required for such isomerization of olefin unit. Without irradiation that is in absence of stimuli, one macrocycles is located at fumaramide station as this is thermodynamically favourable site among other sites and the second macrocycle resides on the methyl-fumaramide station which has second hierarchy in terms of thermodynamics. Upon irradiation of the double bond, isomerization of the olefin part located at station A of the large macrocyclic ring takes place. This weakens hydrogen-bonds between smaller macrocycle located at the region of amide with the large macrocycle. The second most favored site is being occupied by the second small macrocyclic ring; the ring positioned at highest affinity site moves out to third favorable position which is a succinimide amide ester group. When irradiation wavelength is adjusted such that photoisomerization at methyl-fumaramide station takes place, macrocycle located at this position moves to the last binding site D. Fumaramide units are converted to the corresponding *cis*-amide unit by three ways. This isomerization can

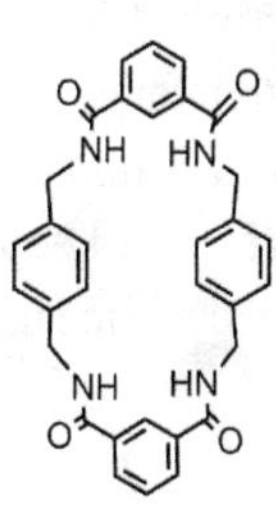

4.19

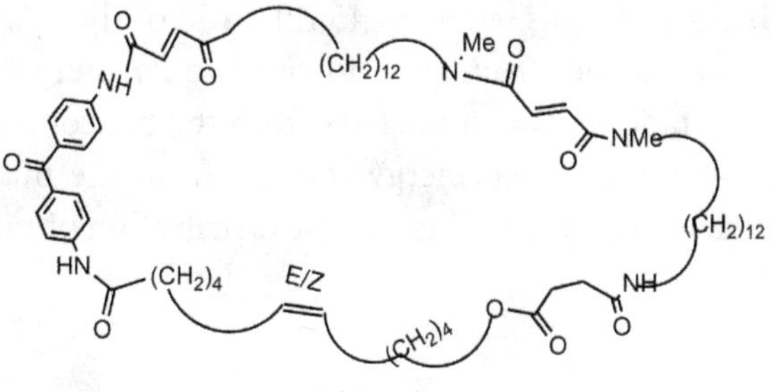

4.20

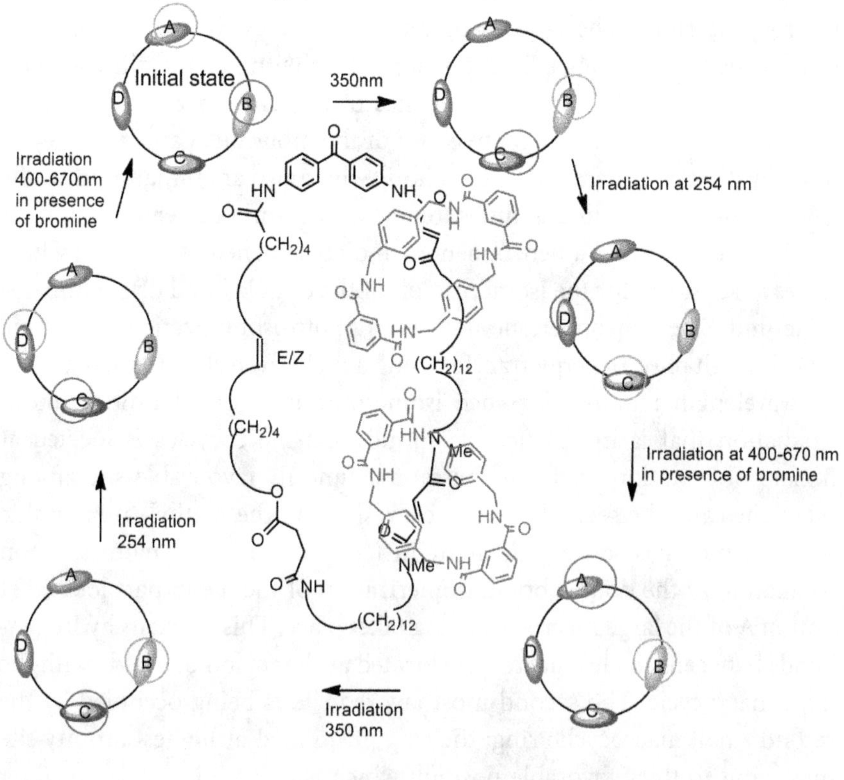

Stimuli guided unidirectional motion

Figure 4.11.   A photochemically driven molecular shuttle.[12b]

be done by three ways. (a) By direct heating, (b) by heating the compound in presence of a catalytic amount of ethylenediamine,(c) by irradiating at 400–670 nm in the presence of a catalytic amount of bromine. The change in position of the rings to different places causes relative motion of the rings over a stipulated path in an opposite directions with respect to the static ring. Thus, by changing the wavelength of irradiation, movement of the rings can be controlled and original states can be regenerated through forward movement through the circular path or by combination of forward and backward motions over equilibrating sites.

4.1.6 *Light driven conformation adjustments*

Hydrogen-bonded network of a tris(N-salicylideneaniline) derivative (Figure 4.12) posses conformational switching property. In this type of molecules structural distortion can be induce by folding and unfolding processes. Electronic properties changes due to change in fold of such a compound. In the example shown in Figure 4.12 has BODIPY

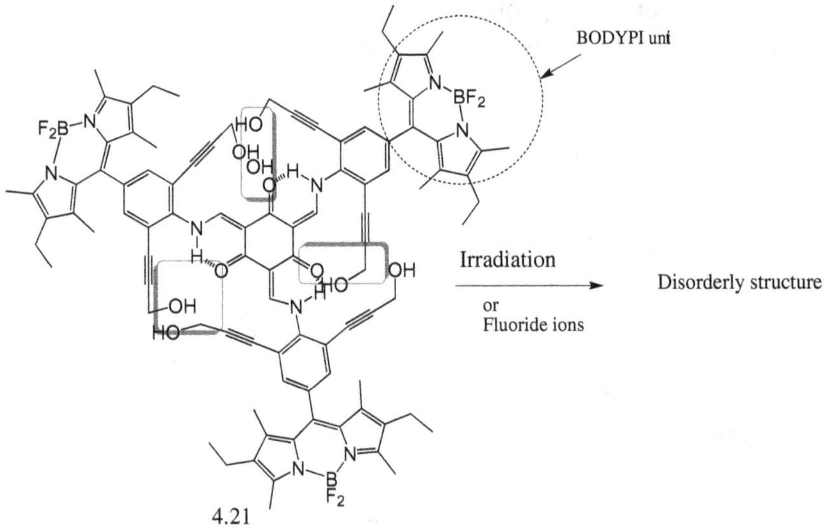

Figure 4.12. Light-driven ordered and disordered structures in a tris(N-salicylideneaniline) derivative.[13]

chromophores which is shown within a circle. The chromophores are also fluorescent and they get activated by light. The molecule adopts a highly ordered hydrogen-bonding array illustrated with rectangles drawn in red color. Energy transfer takes place from the core containing the tris(N-salicylideneaniline) part to the BODIPY units. Basic anion such as fluoride ions interacts with hydrogen bond and makes the molecule unfolded. Due to unfolding and loss of ordered structure there is a substantial decrease in the fluorescence intensity of the BODIPY units with respect to the ordered molecule. Because of highly symmetric nature of structure conformational change is highly cooperative. The ordered structures from disordered structure caused by fluoride ions can be obtained upon interaction with trimethylsilyl chloride.

4.1.7 *Fluorescence modulations*

Equilibrium shown in Figure 4.13 is pushed forward or backward by irradiation with 465 nm followed by irradiation with 614 nm. Photocyclisation reactions not only change the chromophore but also change the geometrical aspect of the molecule. This equilibrium is used in a [2]pseudorotaxane to show switching of photochromic properties using a luminescent lanthanide complex tethered to a crown ether. The diarylperfluoro

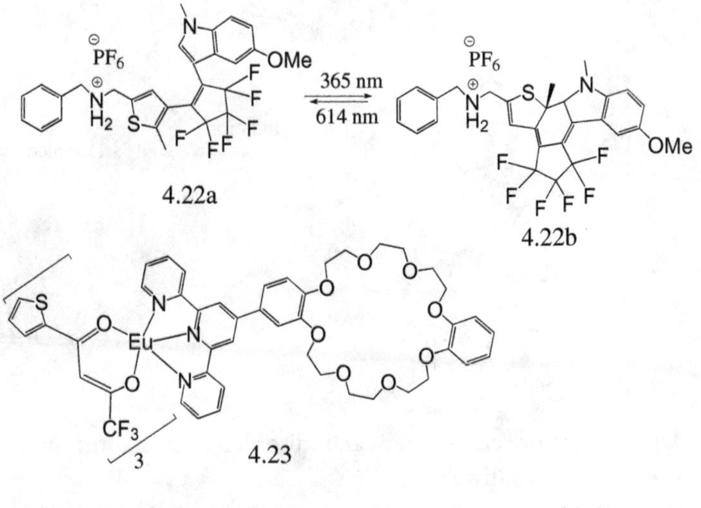

Figure 4.13.   Fluorescence modulation of europium complex in a rotaxane.[14]

cyclopentene is attached to benzyl ammonium cation. This cation recognises crown ether and adopts different orientations based on the cyclic or open form of the axle. Eu^{3+} complex of terpyridinyldibenzo-24-crown-8 (**4.23**) forms complexes reversibly by interacting with ammonium cation. Intramolecular energy transfer of Eu^{3+}ion shows fluorescence emission at 619 nm. The crown ether-ammonium interaction, shows a slight quenching of fluorescence at 619 nm. But there is poor spectral overlap between the donor emission and acceptor absorbance. Under this condition resonance energy transfer (RET) process takes place. When closed form of the compound is irradiated at UV region, spectral overlap increases and fluorescence is quenched by 80%. This pseudorotaxane is reversibly disassembled by interaction with externally added potassium cations and regenerates by the addition of 18-crown-6. This effect happens to due to potassium ion binding to the crown attached to the europium complex. 18-Crown-6 has high affinity for cation which helps it to abstract the potassium ion bound to the complex cation.

Dumb-bell shaped molecule **4.24** is constituted by 4-amino 1,8-naphthalimide fluorophores along with olefinic part as well as azo link. When this molecule is non-covalently interlocked with α-cyclodextrin, it constitutes a rotaxane **4.25a** which shows interesting fluorescence modulation properties. In this rotaxane steric photo-isomerization can cause *cis-trans* isomerisation at N=N or C=C sites, giving different shapes to the molecule. Thus, irradiation with appropriately chosen wavelengths results in four different isomeric structures illustrated in Figure 4.14. *Cis*-form of stilbene upon photo-isomerisation to *trans*-form provides steric hindrance by providing a new geometry to the rotaxane. This facilitates movement of cyclodextrin ring along the thread. To do so the molecule undergoes rotation or vibration, affects the fluorescence emission of the rotaxane. When rotational and vibrational freedom are higher the non-radiative processes are more. Such a process reduces quantum yield which deceases the intensity of emission. In the rotaxane **4.25a** the movement caused by *trans* to *cis* isomer results in restricted motion of the cyclodextrin along the thread of the molecule, due to such restriction intensity of fluorescence enhances by 46%. A similar effect also is seen on the isomerization of azo group, in this case also *cis* form shows higher intensity of fluorescence due to restricted motion. Since the two fluorophores present at two end of the rotaxane are different; the effect caused by changes of *trans-cis* isomerization at two

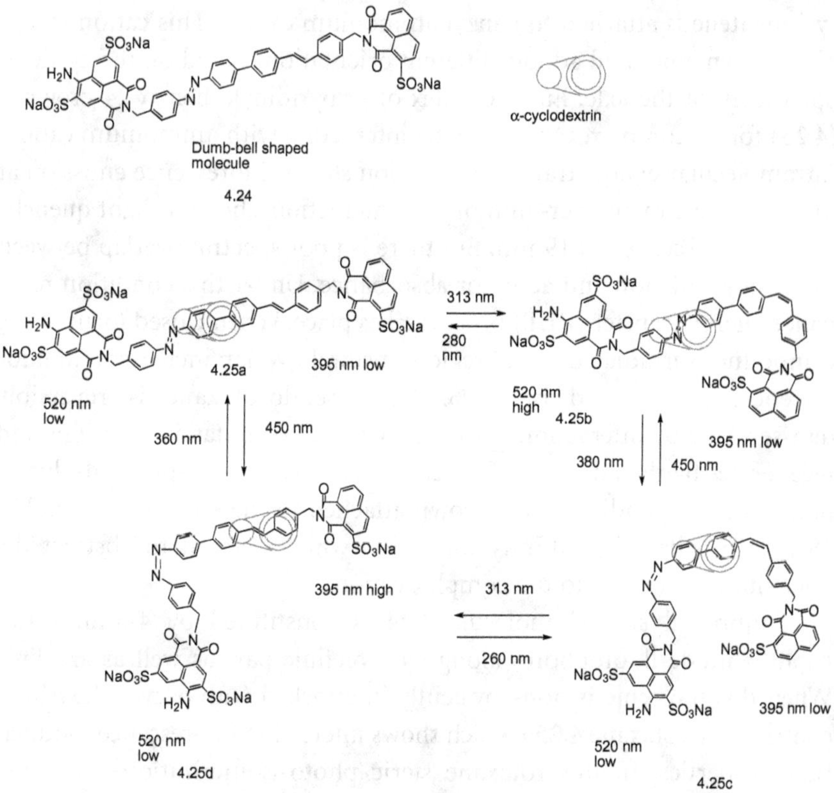

Figure 4.14. A logic gate based on rotaxane through photoisomerisation.[15]

independent sites are dependent on the distance of each fluorophore from the double bond where geometrical changes takes place. The changes fluorescence caused on the stilbene by changing from *trans* to *cis* form on a particular fluorophore present at an end is different from the effect caused by *trans* to *cis* isomerization of the azo group. Thus, a logic gate is constituted by combining such process by changing the course of irradiation to show the performance of the gate in terms of binary addition using two inputs and two outputs. Fluorescence changes in this rotaxane are related to the submolecular motion. The increase or decrease in the fluorescence intensities are reporters of such motion.

4.1.8 *Light-powered molecular pedal*

Adduct **4.26a** shows scissor-like conformational changes which is brought about by suitable light source as an external stimulus. Irradiation cause mechanical twisting of non-covalently linked by Quinoline guest which adopts *syn* and *anti*-forms as shown in equation *a* of Figure 4.15. The host unit of the adduct has a ferrocene moiety with an azobenzene linkage at two different locations. The ferrocene unit acts as pivot which changes the orientation of cyclopentadienyl ring on conversion of the *trans*-azo unit to *cis*-azo unit. This change also alter the positions of the porphyrine rings, movement is such that it effects the orientation of the zinc ion bound to the porphyrin rings. Accordingly the bridging biquinoline ligand to zinc takes place through anti-form of the biquinoline shown in equation b of Figure 4.15. Thus, motion caused in this adduct by irradiation resembles paddle-like motion centering at the ferrocene pivot. This process can be monitored by circular dichroism spectroscopy, where the enantiomers of two states showed characteristic mirror-image CD-bands of one another at 230–370 nm and at 380–440 nm. Thus changes caused by paddle motion relating ferrocene part is

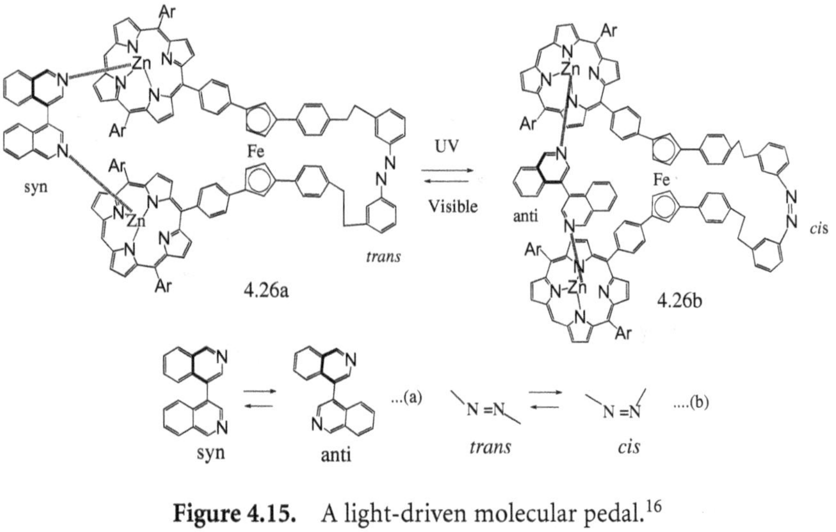

Figure 4.15. A light-driven molecular pedal.[16]

reflected in the region 230–370 nm, whereas that of porphyrin appears in the region of 380–440 nm.

4.1.9 *Helical deformation in photosensitive liquid crystalline materials*

Helical deformations in artificial systems are utilized to produce mechanical work by molecular movement. Such processes caused in liquid-crystal polymer network where right-handed or left-handed macroscopic helices are present at room temperature. Orderly orientation undergoes deformation and mechanical strain changes in such polymeric materials upon irradiation. Due to strong coupling between helical molecules causes changes in orientations of polymeric molecules. Azobenzene containing spiral molecules undergo light responsive reversible changes and they act as nano-scale energy converters. Once they are incorporated into the polymer network through polymerizable acrylate groups in which about 10% azobenzene unit causes deformation of the system upon irradiation (Figure 4.16). Addition of chiral dopants **4.28a** or **4.28b**, induces a left-handed and right-handed twist in the liquid crystal while irradiation of the liquid crystals of polymer **4.27a** respectively.

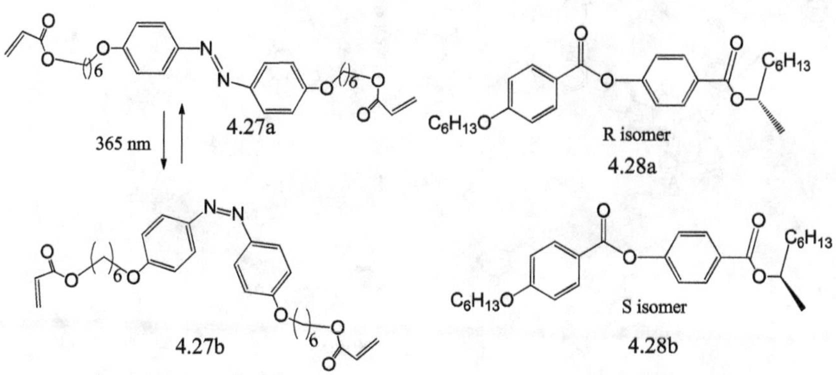

Figure 4.16.   Photosensitive liquid crystalline materials and chiral dopants.[17]

4.2 Electrochemically Driven Machines

4.2.1 *Rotations by redox couple*

Electrical energy is utilised to cause mechanical motions in large number of molecular systems. In such examples the inorganic or organic redox couples are used for causing changes. The reversible couples most often useful for such purposes. In case of inorganic complexes redox reactions changes the oxidation state to modify coordination environment. Redox couple of Nickel is utilised in carborane complexes (Figure 4.17). The complex undergoes oxidation causing the *cis* disposition of the two carborane rings to *trans*-geometry. The process is reversible; hence changes is used in back and forth manner by application of switching voltages.

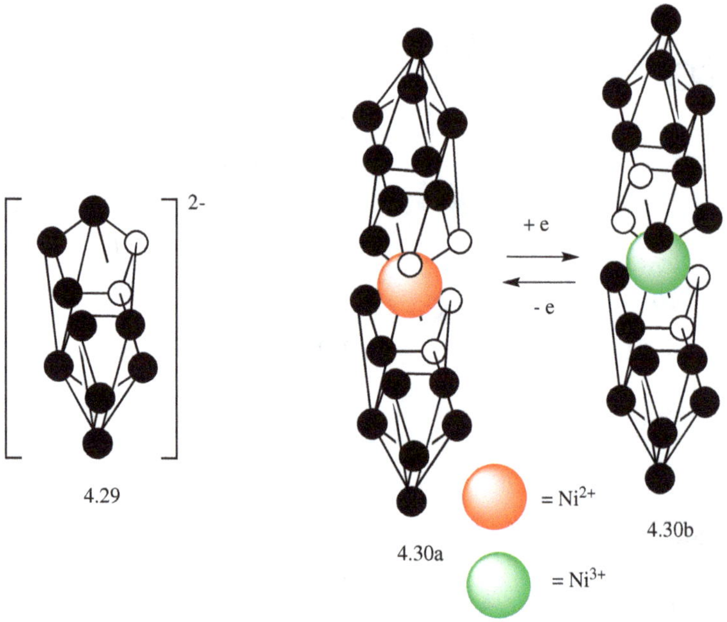

Figure 4.17. Inter-conversion between *cis-trans* form of a nickel-carborane complex.[18]

Copper (II) complexes generally favor distorted octahedron or square pyramid geometry, whereas a complex with copper(I) ion prefers to form tetrahedral geometry. Thus by oxidizing copper center of a tetrahedral copper(I) complex formed by a rotaxane changes geometry. Such a change causes switching of coordination, generating two states of rotaxane **4.31a** (Figure 4.18).

Anion exchange between two metal sites located at two different parts of a molecule can be performed by electrical stimuli. In such redox process affinity for an ion or ligand changes with respect to the original oxidation state. It facilitates transfer of an ion from one metal site to another. The mixed metal complex has a tripodal ligand binding to copper and anchoring a chloride ion at one site and another site is bound to a nickel ion by a cyclam (Figure 4.19). Nickel +3 oxidation state has higher affinity has

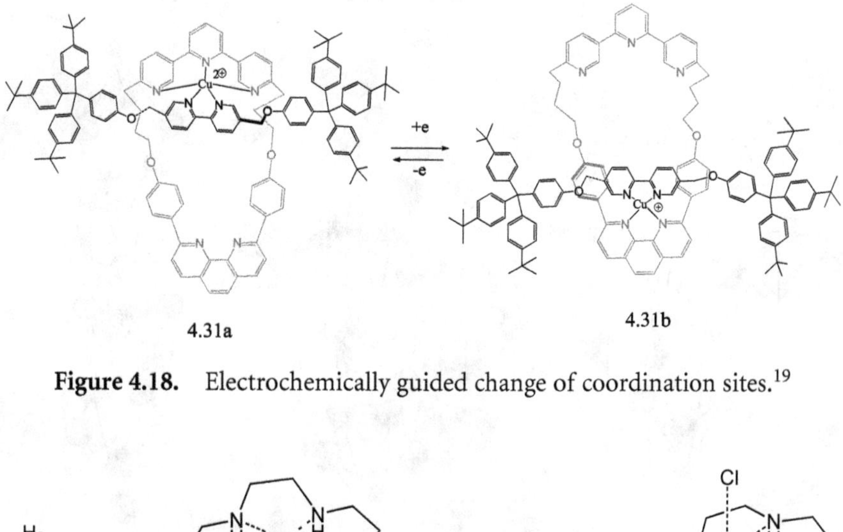

4.31a   4.31b

Figure 4.18. Electrochemically guided change of coordination sites.[19]

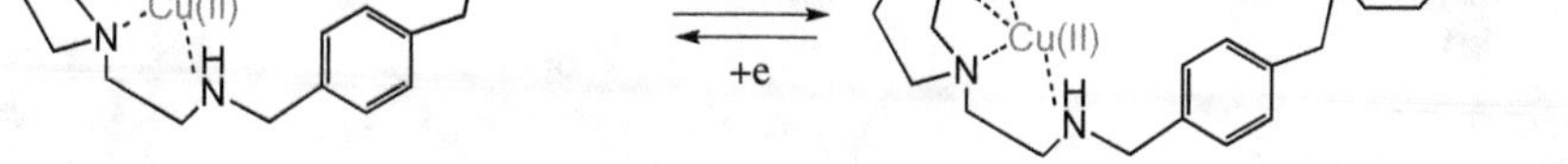

4.32a   4.32b

Figure 4.19. Reversible anion exchange through an electrically driven process.[20]

higher affinity for chloride ion. Such change in oxidation state facilitate shift of chloride ion from the copper ion to the nickel ion. The process can be reversibly done by oxidising and reducing nickel ion by appropriate voltage.

4.2.2 *Machine action due to radicals*

Redox properties of an organic compound can be used to change the binding schemes between interlocked molecules to cause shuttling motion. 1,8-Naphthalimide is a redox active molecule; it undergoes one electron reduction to adopt anionic radical. Thus the anion radical formation helps in exchanging hydrogen bond affinity of 1,8-naphthalimide unit. This principle is used to shuttle movement of a hydrogen bonded ring in a rotaxane. One such example is illustrated in Figure 4.20. In this example amide containing ring prefers to bind to succinimide site. But once an anion radical is generated on the naphthalimide ring the amide bearing ring moves to the end where naphthalimide is located.

Similar redox switches showing colour upon formation of charge-transfer comlex is depicted in Figure 4.21. In this rotaxane the tetracationic bipyridinium derivative is placed over fulvalin part of the axle (Figure 4.21). This part of the rotaxane is easily oxidised by applying voltage. Oxidation generate cationic radical, which forces the cationic ring to move to a position where the naphthalene ether part is located. Such a change generates a new charge-transfer complex. The color of complex **4.34a** in ground state is green whereas upon oxidation, rotaxane **4.34b** shows purple color. The end phenolic groups is utilised to add thiol

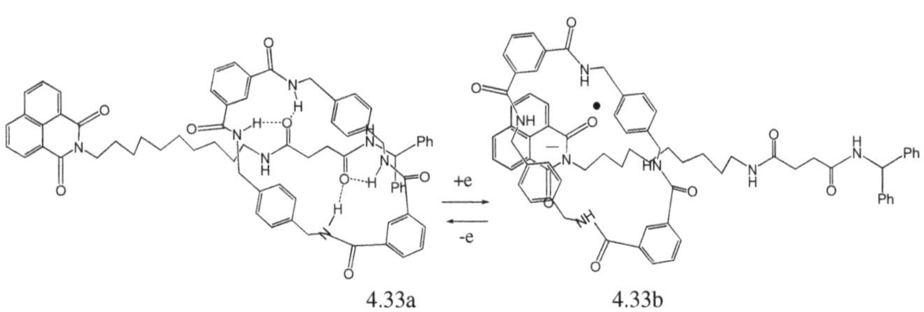

4.33a   4.33b

Figure 4.20. A redox based organic shuttle.[21]

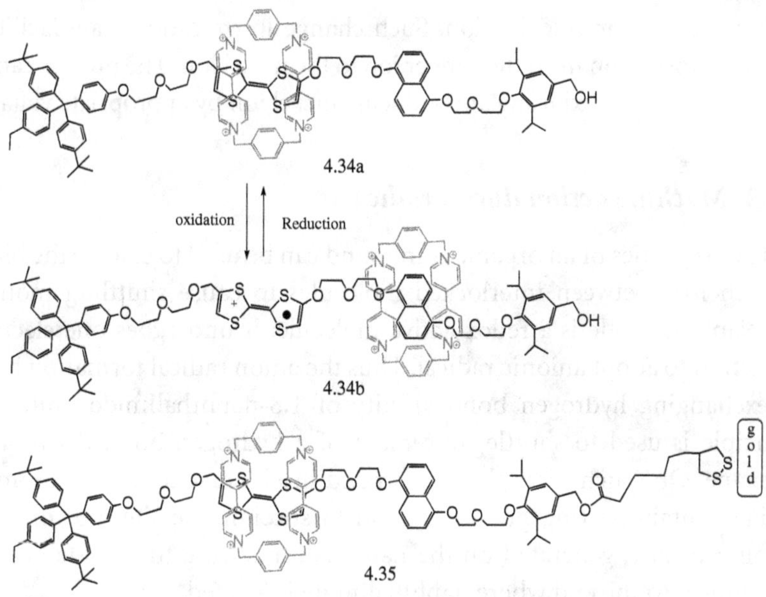

Figure 4.21. An electrochromic device.[22]

containing unit for utilisation in surface modification. In the compound **4.35** a sulphur containing ring is anchored at one end which gets anchored to gold surface makes it a device.

Bistable [3]rotaxane **4.36** derived from an axle **4.37** having thiafulvalene as well as naphthalene ether unit as stations with tetracationic 4,4 -bipyridinium units containing ring behaves like artificial molecular muscles (Figure 4.22). An alternating oxidizing and reducing potentials appropriate to oxidise and reduce the thiofulvalene causes reversible deflections of the rotaxane. Advantage of this property is taken to prepare micro-cantilever devices. For such a purpose the rotaxane is precoated on surface with a gold film. The presence of disulphide unit at the each end of the rotaxane helps in binding to gold. The molecule such as **4.38** alone or 1-dodecanethiol is not useful for cantilever action supporting the necessity of the ring movement over different sites on the axle. The cantilever shows unidirectional deflection when thiofulvalene is oxidized.

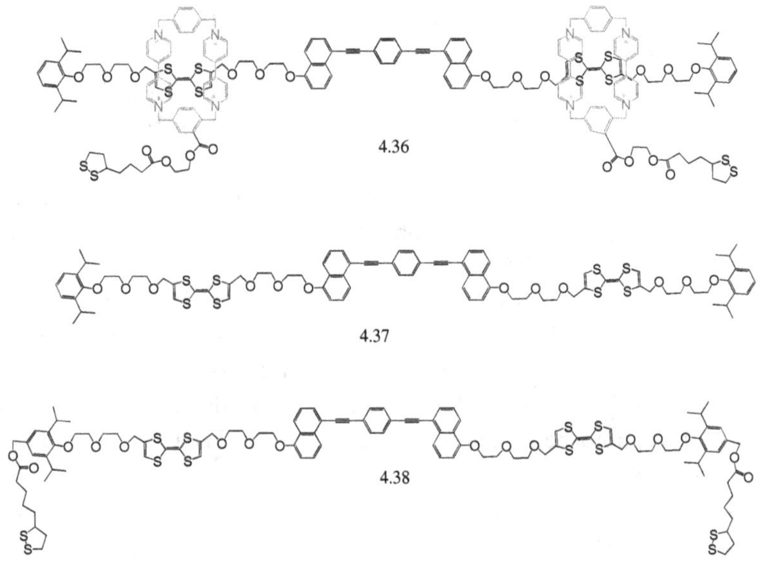

4.36

4.37

4.38

Figure 4.22. Precursor for a micro-cantilever.[23]

References

1. Silvi S, Venturi M, Credi, A (2011) Light operated molecular machines *Chem. Commun.* **47**: 2483–2489.
2. Credi A, Margherita Venturi M (2008) Molecular machines operated by light. *Central Eur. J. Chem.* **6**: 325–339
3. Tian H, Wang Q-C (2006) Recent progress on switchable rotaxanes. *Chem. Soc. Rev.* **35**: 361–374.
4. Shinkai S, Nakaji T, Ogawa T, Shigematsu K, Manabe O (1981) Photoresponsive crown ethers: Photo control of ion extraction and ion-transport by a bis(crown ether) with a bnutterfly-like motion. *J. Am. Chem. Soc.* **103**: 111–115.
5. Koumura NEM, Meetsma GA, Feringa BL (2000) Light-driven molecular rotor: Unidirectional rotation controlled by a single stereogenic center. *J. Am. Chem. Soc.* **122**: 12005–12006.
6. Jager WF, de Jong JC, de Lange B, Huck NPM, Meetsma A, Feringa BL (1995) A highly stereoselective optical switching process based on donor-acceptor-substituted dissymmetric alkenes. *Angew. Chem. Int. Ed.* **34**: 348–350.

7. Koumura N, Zijlstra RWJ, van Delden RA, Harada N, Feringa BL (1999) Light-driven mono directional molecular rotor. *Nature* **401**: 152–155.

8. Murakami H, Kawabuchi A, Kotoo K, Kunitake M, Nakashima N (1997) A light driven molecular shuttle based on a rotaxane. *J. Am. Chem. Soc.* **119**: 7605–7606.

9. Stanier CA, Alderman SJ, Claridge TDW, Anderson HL (2002) Unidirectional photoinduced shuttling in a rotaxane with a symmetric stilbene dumbbell. *Angew. Chem. Int. Ed.* **41**: 1769–1772.

10. Abraham W, Grubert L, Grummt UW, Buck K (2004) A photoswitchable rotaxane with a folded molecular thread. *Chem. Eur. J.* **10**: 3562–3568.

11. Ashton PR, Ballardini R, Balzani V, Credi A, Dress R, Ishow E, Kleverlaan CJ, Kocian O, Preece JA, Spencer N, Stoddart JF, Venturi M, Wenger S (2000) A photochemically driven molecular-level abacus. *Chem. Eur. J.* **6**: 3558–3574.

12a. Serreli V, Lee CF, Kay ER, Leigh DA (2007) A molecular information ratchet. *Nature* **445**, 523–527.

12b. Leigh DA, Wong JKY, Dehez F.; Zerbetto, F (2013) Unidirectional rotation in a mechanically interlocked molecular rotor. *Nature* **424**: 174–179.

13. Riddle JA, Jiang X, Huffman J, Lee D (2007) Signal-amplifying resonance energy transfer: A dynamic multichromophore array for allosteric switching. *Angew. Chem. Int. Ed.* **46**: 7019–7022.

14. Cheng, H. B.; Zhang, H. Y.; Liu, Y. (2013) Dual-stimulus luminescent lanthanide molecular switch based on an unsymmetrical diarylperfluorocyclopentene. *J. Am. Chem. Soc.* **135**: 10190–10193.

15. Qu DH, Wang QC, Tian H (2005) A half adder based on a photochemically driven [2]rotaxane. *Angew. Chem. Int. Ed.* **44**: 5296–5299.

16. Muraoka T, Kinbara K, Aida T (2006) Mechanical twisting of a guest by a photo responsive host. *Nature* **440**: 512–515.

17. Huck NPM, Feringa BL, vanDoren H (1995) Chiroptical switching between liquid crystalline phases. *J. Am. Chem. Soc.* **117**: 9929–9930.

18. Hawthorne MF, Zink JI, Skelton JM, Bayer MJ, Liu C, Livshits E, Baer R, Neuhauser D (2004) Electrical or photo control of the rotary motion of a metallacarborane. *Science* **303**: 1849–1851.

19. Feringa BL, van Delden RA, Koumura N, Geertsema E (2000) Chiroptical molecular switches. *Chem. Rev.* **100**: 1789–1816.

20. Fabbrizzi L, Gatti F, Pallavicini P, Zambarbieri E (1999) Redox driven intramolecular anion translocation between transition metal centers. *Chem. Eur. J.* **5**: 682–690.

21. Altieri A, Gatti FG, Kay ER, Leigh DA, Martel D, Paolucci F, Slawin AMZ, Wong JKY (2003) Electrochemically switchable hydrogen-bonded molecular shuttles *J. Am. Chem. Soc.* **125**: 8644–8654.

22. Tseng H-R, Vigon SA, Stoddart FA (2003) Towards chemically controlled molecular machinery. *Angew. Chem. Int.. Ed.* **42**: 1491–1495.

23. Liu Y, Flood AH, Bonvallett PA, Vignon SA, Northrop BH, Tseng HR, Jeppesen JO, Huang TJ., Brough B, Baller M, Magonov S, Solares SD, Goddard WA III, Ho CM, Stoddart JF (2005) Linear artificial molecular muscles. *J. Am. Chem. Soc.* **127**: 9745–9759.

5

Artificial Molecular Machine Based on DNA

5.1 Principles

Complementary hydrogen bonds between the heterocyclic bases such as adenine, cytosine, guanine and thiamine of deoxyribonuclic acid (DNA) are amenable for modifications by competitive assembling, disassembling process. DNA has ability to remain as single strand or form duplex structure. Challenging aspects of DNA nanostructured assembly formed by interactions between specific base-pair are to utilize them to fabricate nano-machines and to is to cause movements in response to external stimuli. DNA nano-machine could be regarded as a kind of molecular machine that is made up of assembled DNA structures integrated with external stimuli responding mechanism.[1] The molecular machine designed based on the principle of assembling single stranded DNA to duplex structure and exchange of strands in duplex structures. Many such processes are controlled by thermodynamics of binding preferences and guided by external stimuli or by chemicals. Light-induced conformation changes of DNA in functionalized DNA with substrates capable of photo-chemical transformation such as *cis-trans* isomerisation causes reversible structural changes[2] causing contraction, expansion or in packing arrangement contributing to the phase related properties.

5.2 DNA Tweezers

Structural modifications on DNA duplex structures can be utilized in designing molecular machine. For example a more stable duplex can be

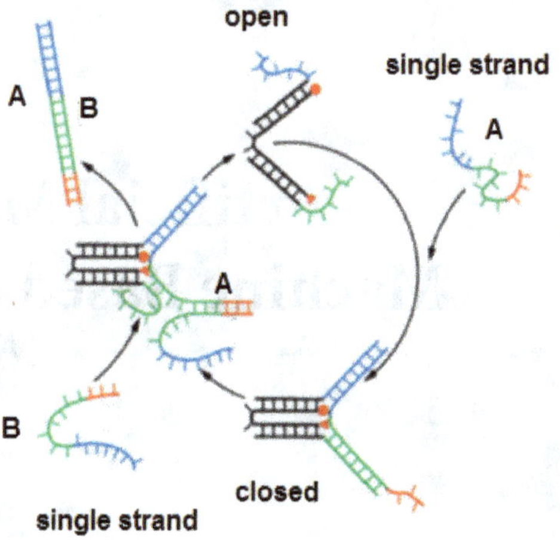

Figure 5.1. DNA tweezers from competitive binding of single strand DNA A and B.

constructed by "chain-exchange reaction" or "strand-exchange reaction" in which short DNA strand is exchanged by a longer strand. Such changes in nanostructured DNA strands induce motions. An example of such a molecular machine is shown in Figure 5.1. This example is a tweezers like nano machine.[3] In this example a hinge holds three single-strands of DNA (A); which are fixed such that an angular geometry is formed by two rigid duplex arms. While assembling of such three DNA single strands, two of them have two dangling ends. When the two ends of the arms are thermo-dynamically separated it is in open state of the tweezers. To convert the open state to close state fourth single strand A is used. This strand forms duplex structure with the dangling ends and pulls the two arms together provide a closed state. The device can be brought to open state again by using another single strand DNA 'B' which has complementary binding ability with strand A. Thus new strand B will tend to form double strand with A which is a lower free energy state. The addition of A and B in alternate manner makes a cycle to operate like a machine. In the process machine produces AB as waste. Here, A and B are fuel for competitive duplex formation but the function of A is just opposite of B hence A is called "fuel" and B is called "anti-fuel".

Disadvantage of this machine is the duplex wastes produced in this example. To improve performance, non-DNA stimuli are used to control motions. For such purposes intercalator, oscillatory reaction, ions, and chelating agents are used, which controls the branched junctions.

5.3 DNA Walker

The simplest unidirectional motion of DNA molecule is linear walking. Pyrene when excited by photo radiation it helps in S-S bond cleavage to form thiol. This reaction along with the propensity of base-pair to form duplex structure is utilized to make walking motion of DNA over a DNA tile.[4] A walking nano-machine illustrated in Figure 5.2 has a supporting rectangular DNA tile having binding sites.

Example shown in Figure 5.2 illustrating walking action has four sites to hold vertically DNA molecules as stator strands. The walking strand chosen among these strands forms duplex with stator strands. The walking strand has two strands connected to each other one is short and another is relatively long, where longer one has an oligonucleotide modified with two pyrene molecules. Stator strand has also two strands which are connected

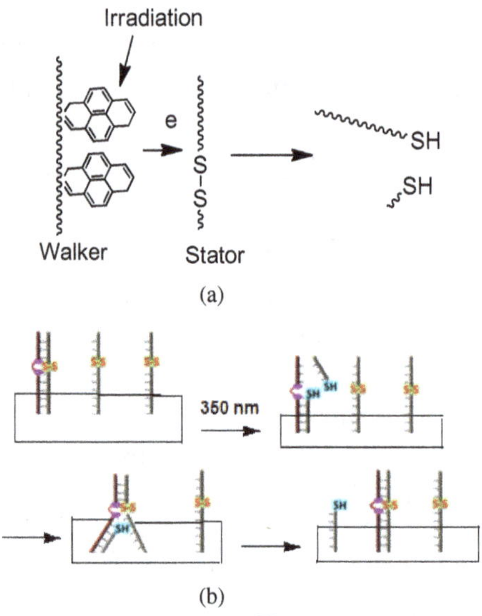

Figure 5.2. (a) Principle followed in a light-driven walking system (b) Mechanism of forward motion of the walker.

by disulfide bond. The static strand has two additional bases. These bases hybridize with the walker DNA strand. This walker is designed as the terminal point of walker. It has three extra thymine bases, which prevents the dissociation of the walker from the stator and to terminate its movement. Stators F_1-F_4 are separated by a distance of ~6 nm linearly paced over the tile. Shorter potions of DNA of four stators help to identify the location of walker on the tile. Long track requires much longer time of irradiation, which might induce photo damage. Pyrene molecules on photo-irradiation at 350 nm cleave disulfide bond to form thiol on a neighbour molecule. The overall forward-step of the walker strand is shown in Figure 5.2b. Thus partially dissociated walker reacts with S-S bond of neighbour stator strand. While doing that double strand is formed between the pyrene containing strand and neighbouring strand gets free. This causes movement of the pyrene containing strand a step forward. The process restarts and step wise forward motion takes place. The unidirectional stepping of the walker continues under continuous photo-irradiation.

5.4 DNA Based Molecular Gear

DNA based molecular gears have capability to reverse the direction of rotation, control speed of rotation and to change rotational motion to a different axis. Such gear[5] comprising of combination of A and B are illustrated in Figure 5.3. Each unit has four DNA single strands. Among which one is a circular strand (C) and three other peripheral linear strands (represented by P). Two units A and B of gears are based on same C strand. Difference is that they have different sets of strands (P). Four strands are in the form of duplex structures over the circular strand. While forming such duplexes, there is residual portion of unhybrid strand portion which act as teeth (T). Gears A and B are independent of each other and do not interact independently. To make an integrated system unit A and unit B are connected by linker strands (L). In order to separate the two units connected by linker L another set of DNA strands are used they are called removal strands (R). During operation two gears are linked at a tooth position by L_1. When L_2 is added another duplex is formed by using two portions of free strands on the circles and the two rings comes closer by making two bridges between A and B. L_1 is complementary strand to R_1,

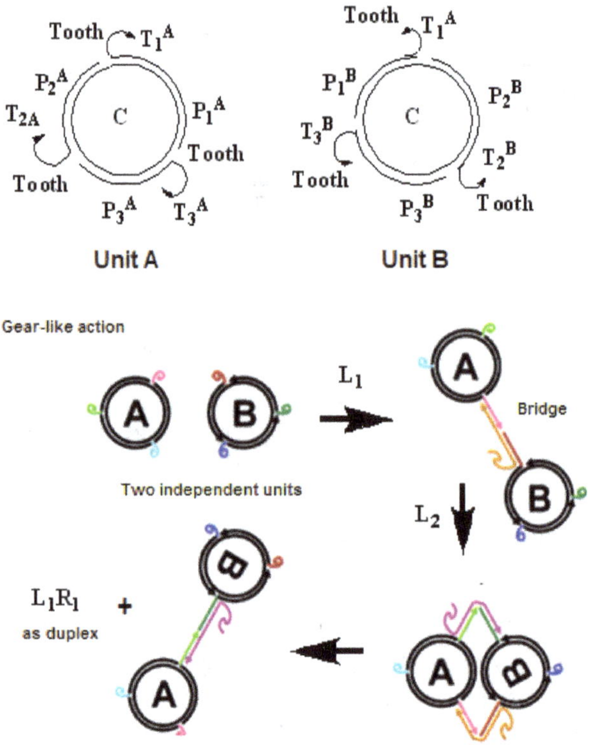

Figure 5.3. DNA Nano-gear design and action.

thus addition of R_1 replaces L_1 to make single bridge between A and B. Thus the balls are separated and due to the presence of portion of single strand as teeth, the rotation as well as directional motion occurs to behave like a gear.

5.5 pH Dependent DNA Nano-machine

Hydrogen ions play important role in every sphere of chemical and biological processes. In self-assembling of DNA base pairs protonation decides a major role. This is due to protonation of a hydrogen acceptor makes it a site for hydrogen bond donor. Thus by protonation or deprotonation of hydrogen bonding sites can change the self-assembling process. This aspect is utilized in construction of DNA based molecular machines.[6]

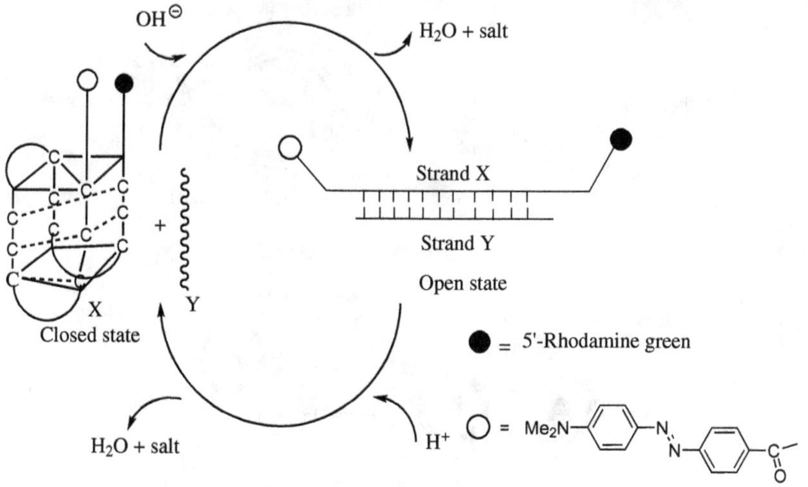

Figure 5.4.   A DNA-based nano-motor driven by pH.

Example given in Figure 5.4 it is shown that acidic pH favors formation of a four-stranded cytosine rich structure having C^+CC triplet. In acid medium as a consequence of formation of cytosinium cation from cytocine base protonation (C^+) extra hydrogen bonds are formed. Example given in Figure 5.4 is an example of DNA 21mer strand 'X' that has four CCC stretches. This strand also has a partial complementary strand Y. At low acid concentration, CCC stretch is partially protonated and it forms intercalated C^+C base pairs. Thus, single strand X folds to form a compact structure. In terms of molecular machine derived from X it is a closed state. By increasing pH slightly the neutral form of the strand is regained and this pH change causes unfolding of X. As a result of regeneration of native state of X state, it forms duplex structure XY and generate a state which is double stranded open state. Thus, by changing pH reversible switching between a compact and an extended state takes place. This machine results salt and water as wastes.

5.6 Molecular Assembler

Programmed molecular machines are needed to assemble and propagate reactions in relating day to day biological activities to regulate activities of life. Such programmed systems operates in biology to act as

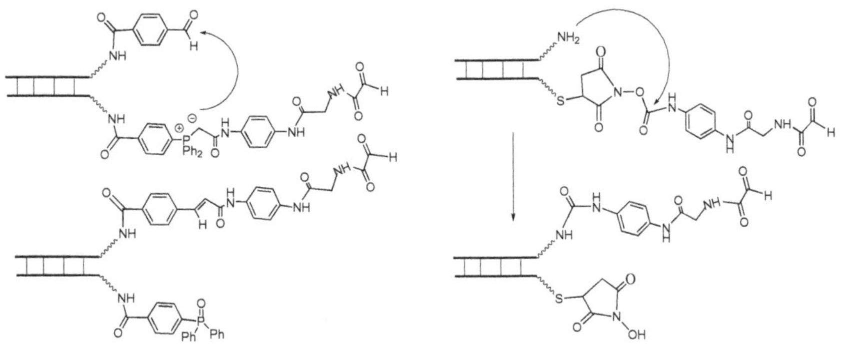

Figure 5.5. Propagation of oligomers by DNA templates; (a) Olefin formation and (b) Peptide bond formation.

sequential synthesis of polymeric molecules in programmed manner to deliver reactant as well as substrate-like carrier to make synthetic molecular machinery.[7] Due to different folds and equilibration of single strand and double strand structures of functionalized DNA, they control formation of covalent bonds. Thus, DNA hybridization reactions are utilized to provide appropriate template and guide building blocks attached linkers such as peptide bonds. In two examples it is shown that the double helix formation providing appropriate template to bring two functional groups to carry out group transfer reaction. In the example shown in Figure 5.5(a) the carbanion of the ylide, forms olefin by reaction with an aldehyde group, thereby transfers a portion of peptide from one end to another. In the example shown in Figure 5.5(b) does a similar group transfer but in this case an ester linkage reacts with an amine group to form new amide bond while transferring a group from one end to another. Such group transfer reactions are translated to assembler program comprising of DNA instruction strands to act as molecular machine. These strands have ability to interact among them in controlled manner to define reaction sequences.

5.7 Trends of Study

Progress and utilization of equipment such as transmission electron microscope, atomic force microscope and various other tools that monitors an event at pico to femto-second have added great impetus to pursue and transform working molecular machine to reality. Interest in bio-molecular machines and robots will remain as challenge for next generation of machines. Developments of new bio-molecular machines similar to the effects created a cellular level such as motion, force, or a signal and utilize them for practical purpose remains a challenge.

References

1. Liu H, Liu D (2009) DNA nano-machines and their functional evolution, *Chem. Commun.* 2625–2636.
2. Kuzyk A, Yang Y, Duan X, Stoll S, Govorov AO, Sugiyama H, Endo M, Liu N (2016) A light-driven three-dimensional plasmonic nano system that translates molecular motion into reversible chiroptical function. *Nature Commun* **7**: 10591.
3. Yurke B, Turberfield AJ, Mills Jr AP, Simmel FC, Neumann JL (2000) A DNA-fuelled molecular machine made of DNA. *Nature* **406**: 605–608.
4. Yang Y, Goetzfried MA, Hidaka K, You M, Tan W, Sugiyama H, Endo M (2015) Direct visualization of walking motions of photocontrolled nanomachine on the DNA nanostructure, *Nano Lett.* **15**: 6672–6676.
5. Tian Y, Mao C (2004) Molecular gears: a pair of DNA circles continuously rolls against each other. *J. Am. Chem. Soc.* **126**: 11410–11411.
6. Liu D, Balasubramanian S (2003) A Proton-Fuelled DNA Nanomachine, *Angew. Chem. Int. Ed.* **42**: 5734–5736.
7. Meng W, Muscat RA, McKee ML, Milnes PJ, El-Sagheer AH, Bath J, Davis BG, Brown T, O'Reilly RK, Turberfield AJ (2016) An autonomous molecular assembler for programmable chemical synthesis. *Nature Chem.* **8**: 543–548.

Index

9789813223707